Encyclopaedia of

World Geography

Encyclopaedia of
World Geography

Vol. - 4

Chandra Vijay Purty

ABD PUBLISHERS
Jaipur ■ New Delhi

ABD PUBLISHERS

Regd. Off. "Bony Residency", Gate No. 2
Opp. Tilak Public School, Vishweswariya Nagar,
Gopalpura Road, Jaipur - 302018, India
Ph: 0141 - 2761280, Telefax: 2761381
Email: abdpublishersjpr@gmail.com

Delhi Office: 102, 1st Floor, Satyam House,
4327/3, Ansari Road, Darya Ganj,
New Delhi - 110002, India
Ph: 011-45652440

visit us: www.oxfordbookcompany.com

Published in India, 2024

ISBN : 9788183763127 (7 Vols. Set)

Typeset by Shivangi Computer, Jaipur

Printed and bound in India by Thomson Press Ltd.,
New Delhi

Preface

Geography is the study of the Earth and its lands, features, inhabitants, and phenomena. A literal translation would be "to describe or write about the Earth". The first person to use the word "geography" was Eratosthenes (276-194 B.C.). Four historical traditions in geographical research are the spatial analysis of natural and human phenomena (geography as a study of distribution), area studies (places and regions), study of man-land relationship, and research in earth sciences. Nonetheless, modern geography is an all-encompassing discipline that foremost seeks to understand the Earth and all of its human and natural complexities—not merely where objects are, but how they have changed and come to be. As "the bridge between the human and physical sciences," geography is divided into two main branches—human geography and physical geography.

The oldest known world maps date back to ancient Babylon from the 9th century BC. The best known Babylonian world map, however, is the *Imago Mundi* of 600 BC. The map as reconstructed by Eckhard Unger shows Babylon on the Euphrates, surrounded by a circular landmass showing Assyria, Urartu and several cities, in turn surrounded by a "bitter river" (Oceanus), with seven islands arranged around it so as to form a seven-pointed star. The accompanying text mentions seven outer regions beyond the encircling ocean. The descriptions of five of them have survived. In contrast to the *Imago Mundi*, an earlier Babylonian world map dating back to the 9th century BC depicted Babylon as being further north from the centre of the world, though it is not certain what that centre was supposed to represent.

The ideas of Anaximander (c. 610 B.C.-c. 545 B.C.), considered by later Greek writers to be the true founder of geography, come to us through fragments quoted by his successors. Anaximander is credited with the invention of the gnomon, the simple yet efficient Greek instrument that allowed the early measurement of latitude. Thales, Anaximander is also credited with the prediction of eclipses. The foundations of geography can be traced to the ancient cultures, such as the ancient, medieval, and early modern Chinese. The Greeks, who were the first to explore geography as both art and science, achieved this through Cartography, Philosophy, and Literature, or through Mathematics. There is some debate about who was the first person to assert that the Earth is spherical in shape, with the credit going either to Parmenides or Pythagoras. Anaxagoras was able to demonstrate that the profile of the Earth was circular by explaining eclipses. However, he still believed that the Earth was a flat disk, as did many of his contemporaries. One of the first estimates of the radius of the Earth was made by Eratosthenes.

The first rigorous system of latitude and longitude lines is credited to Hipparchus. He employed a sexagesimal system that was derived from Babylonian mathematics. The parallels and meridians were sub-divided into 360°, with each degree further subdivided 602 (minutes). To measure the longitude at different location on Earth, he suggested using eclipses to determine the relative difference in time. The extensive mapping by the Romans as they explored new lands would later provide a high level of information for Ptolemy to construct detailed atlases. He extended the work of Hipparchus, using a grid system on his maps and adopting a length of 56.5 miles for a degree.

From the 3rd century onwards, Chinese methods of geographical study and writing of geographical literature became much more complex than what was found in Europe at the time (until the 13th century). Chinese geographers such as Liu An, Pei Xiu, Jia Dan, Shen Kuo, Fan Chengda, Zhou Daguan, and Xu Xiake wrote important treatises, yet by the 17th century, advanced ideas and methods of Western-style geography were adopted in China.

This book contains the fundamental and basic information of the subject.

Chandra Vijay Purty

Contents

1

The Middle Atlantic States—Leaders in Industry and Trade

The Mid-Atlantic states, also called middle Atlantic states or simply the mid Atlantic, form a region of the United States generally located between New England and the South. Its exact definition differs upon source, but the region often includes Delaware, Maryland, New Jersey, Pennsylvania, Washington D.C., and sometimes New York, Virginia and West Virginia.

The Mid-Atlantic has played an important role in the development of American culture, commerce, trade, and industry, yet it is one of the least self-conscious of American regions. It has been called "the typically American" region by Frederick Jackson Turner. Religious pluralism and ethnic diversity have been important elements of Mid-Atlantic society from its settlement by Dutch, Swedes, English Catholics, and Quakers through to the period of English rule, and beyond. After the American Revolution, the Mid-Atlantic region hosted each of the historic capitals of the United States, including the current federal capital, Washington D.C.

In the early part of the nineteenth century, New York and Pennsylvania overtook Virginia as the most populous states and the New England states as the country's most important trading and industrial centres. Large numbers of German, Irish, Italian, Jewish, Polish, and other immigrants transformed the region, especially coastal cities such as New York City,

Newark, Philadelphia, and Baltimore, but also interior cities such as Pittsburgh and Buffalo.

New York City, with its skyscrapers, subways, and headquarters of the United Nations, emerged in the twentieth century as an icon of modernity and American economic and cultural power. It suffered the brunt of the September 11 attacks, along with two other places in the Mid-Atlantic, Arlington, Virginia and Shanksville, Pennsylvania. By the twenty-first century, the coastal areas of the Mid-Atlantic were thoroughly urbanized. The Northeast Corridor and Interstate 95 linked an almost contiguous sprawl of suburbs and large and small cities, forming the Mid-Atlantic portion of the northeast megalopolis, one of the world's most important concentrations of finance, media, communications, education, medicine, and technology. Waves of Chinese, Indian, Mexican, Russian, Dominican, Jamaican, Filipinos, Pakistani, Guatemalan, Salvadoran, and other immigrants now are further transforming the Mid-Atlantic economy and culture.

The Mid-Atlantic is a relatively affluent region of the nation, having 43 of the 100 highest-income counties in the nation based on median household income and 33 of the top 100 based on per capita income. Most of the Mid-Atlantic states rank among the 15 highest-income states in the nation by median household income and per capita income.

Defining the Mid-Atlantic

There are differing interpretations as to the composition of the Mid-Atlantic. Sometimes, the nucleus is considered to consist of Maryland, Delaware, and Virginia, with additional states possibly included. Other sources consider New Jersey, New York, and Pennsylvania to be the core Mid-Atlantic states, with others sometimes included. For example, since the 1910 census, the Mid-Atlantic Census Division has included New Jersey, New York, and Pennsylvania, which combined with the New England Division, comprised the Northeast Census Region.

Virginia and West Virginia are atypical of this region in several ways. They are the only states to lie primarily within the southern American dialect region, and the major religious tradition in both states is Evangelical Christian, 31% in Virginia and 36% in West Virginia. Although a few of West Virginia's eastern panhandle counties are considered part of the Washington, D.C. MSA, the major portion of the state is rural, and there are no major or even large cities.

The 'Typically American' Region

Frederick Jackson Turner wrote in 1893 about the important role the Mid-Atlantic or "Middle region" had played in the formation of the national American culture, and defined it as "the typical American region".

> *"The Middle region, entered by New York harbour, was an open door to all Europe. The tide-water part of the South represented typical Englishmen, modified by a warm climate and servile labour, and living in baronial fashion on great plantations; New England stood for a special English movement— Puritanism. The Middle region was less English than the other sections. It had a wide mixture of nationalities, a varied society, the mixed town and county system of local government, a varied economic life, many religious sects. In short, it was a region mediating between New England and the South, and the East and the West. It represented that composite nationality which the contemporary United States exhibits, that juxtaposition of non-English groups, occupying a valley or a little settlement, and presenting reflections of the map of Europe in their variety. It was democratic and nonsectional, if not national; "easy, tolerant, and contented;" rooted strongly in material prosperity. It was typical of the modern United States. It was least sectional, not only because it lay between North and South, but also because with no barriers to shut out its frontiers from its settled region, and with a system of connecting waterways, the Middle region mediated between East and West as well as between North and South. Thus it became the typically American region. Even the New Englander, who was shut out from the frontier by the Middle region, tarrying in New York or Pennsylvania on his westward march, lost the acuteness of his sectionalism on the way."*
>
> —*The Frontier in American History*

History

Shipping and trade have been important to the Mid-Atlantic economy since the beginning of the colonial era.

The explorer Giovanni da Verrazzano was the first European since the unenduring Norse colonization of the Americas to have explored the Atlantic coast, travelling from what is now called The Carolinas to Newfoundland. He entered what is now called Lower New York Bay in 1524, greeted by what are presumed to have been the Lenape in canoes, the Native American people who long inhabited that area. Henry Hudson later extensively explored that region in 1609 and claimed it for the Dutch, who then created a fur-trading post in Albany in 1614. The Jamestown Settlement was the first permanent English colony in North America seven years earlier in 1607, founded in what is now called Virginia.

From early colonial times, the Mid-Atlantic region was settled by a wider range of European people than in New England or the South. The Dutch New Netherland settlement along the Hudson River in New York and New Jersey, and for a time, New Sweden along the Delaware River in Delaware, divided the two great bulwarks of English settlement from each other.

The original English settlements in the region notably provided refuge to religious minorities, Maryland to Roman Catholics, and Pennsylvania to Quakers and the mostly Anabaptist Pennsylvania Dutch. In time, all these settlements fell under English colonial control, but the region continued to be a magnet for people of diverse nationalities.

The area that came to be known as the Middle Colonies served as a strategic bridge between the North and South. The New York and New Jersey campaign during the American Revolutionary War saw more battles than any other theater of the conflict. Philadelphia, midway between the northern and southern colonies, was home to the Continental Congress, the convention of delegates who organized the American Revolution. The same city was the birthplace of the Declaration of Independence in 1776 and the United States Constitution in 1787, while the United States Bill of Rights was drafted and ratified, and the first Supreme Court of the United States sat for the first time, in the first capital under the Constitution at New York City.

While early settlers were mostly farmers, traders, and fishermen, the Mid-Atlantic states provided the young United States with heavy industry and served as the "melting pot" of new immigrants from Europe. Cities grew along major ports, shipping routes, and waterways. Such flourishing cities included New York City and Newark on opposite sides of the Hudson River, Philadelphia on the Delaware River, and Baltimore on the Chesapeake Bay.

Cities and Urban Areas

Combined Statistical Areas

Combined statistical areas with more than 1,000,000 people:

- New York-Newark-Bridgeport, NY-NJ-CT-PA CSA
- Washington-Baltimore-Northern Virginia, DC-MD-VA-WV CSA
- Philadelphia-Camden-Vineland, PA-NJ-DE-MD CSA
- Pittsburgh-New Castle, PA CSA
- Buffalo-Niagara-Cattaraugus, NY CSA
- Albany-Schenectady-Amsterdam, NY CSA
- Rochester-Batavia-Seneca Falls, NY CSA.

Large Cities

Cities with more than 200,000 people:

- Arlington
- Baltimore
- Buffalo
- Chesapeake
- Jersey City
- New York City
- Newark
- Norfolk
- Pittsburgh
- Philadelphia
- Richmond
- Rochester

- Virginia Beach
- Washington, D.C.

State capitals:

- Albany, New York
- Annapolis, Maryland
- Charleston, West Virginia
- Dover, Delaware
- Harrisburg, Pennsylvania
- Richmond, Virginia
- Trenton, New Jersey.

The Southern States— A New Industrial Frontier

The Southern United States—commonly referred to as the American South, Dixie, Down South, or simply the South—constitutes a large distinctive region in the southeastern and south-central United States.

Because of the region's unique cultural and historic heritage, including Native Americans, early European settlements of English, Ulster Scots, Scottish, and German heritage, importation of hundreds of thousands of enslaved Africans, growth of a large proportion of African Americans in the population, reliance on slave labour, and legacy of the Confederacy after the American Civil War, the South developed its own customs, literature, musical styles, and varied cuisines that have profoundly shaped traditional American culture.

In the last few decades, the South has become more industrialized and urban, attracting numerous national and international migrants. The American South is among the fastest-growing areas in the United States. Despite rapid economic growth, the South still has persistent poverty, and every Southern state with the exceptions of Virginia and Florida has a higher poverty rate than the American average. Poverty is especially prevalent in rural areas.

Geography

As defined by the United States Census Bureau, the Southern region of the United States includes sixteen states. Thirty-six percent of all U.S. Residents lived in the South, the nation's most populous region. The Census Bureau defined three smaller units, or divisions:

- The South Atlantic States: Florida, Georgia, Maryland, North Carolina, South Carolina, Virginia, West Virginia, and Delaware
- The East South Central States: Alabama, Kentucky, Mississippi and Tennessee
- The West South Central States: Arkansas, Louisiana, Oklahoma, and Texas.

Other terms related to the South include:

- The Old South: usually the original Southern colonies: Virginia, Delaware, Maryland, Georgia, North Carolina, and South Carolina.
- The New South: usually including the South Atlantic States.
- The Solid South: region controlled by the U.S. Democratic Party from 1877 to 1964. Includes at least all the 11 former Confederate States.
- Southern Appalachia: mainly refers to areas situated in the Southern Appalachian Mountains, namely Eastern Kentucky, East Tennessee, Western North Carolina, Western Maryland, West Virginia, Southwest Virginia, North Georgia, and Northwestern South Carolina.
- Southeastern United States: usually including the Carolinas, Virginia, Tennessee, Kentucky, West Virginia, Georgia, Alabama, Mississippi, and Florida
- The Deep South: various definitions, usually including Louisiana, Alabama, Mississippi, Georgia, and South Carolina. Occasionally, parts of adjoining states are included (sections of East Texas, delta areas of Arkansas and Tennessee, and parts of Florida such as the Panhandle and the north-central part of the state).
- The Gulf South: various definitions, usually including Gulf coasts of Florida, Louisiana, Mississippi, Texas and Alabama.
- The Upper South: Kentucky, Virginia, West Virginia, Tennessee, and North Carolina.

- Dixie: various definitions, but most commonly associated with the 11 states of the Old Confederacy.
- The Mid-South: defined by the Census as the South Central United States; in another informal definition, Tennessee, Arkansas, Mississippi, and sometimes adjoining areas of other states.
- Border South: Missouri, Kentucky, Maryland, and Delaware were states on the outer rim of the Confederacy that did not secede from the United States but did have significant numbers of residents who joined the Confederate armed forces. Kentucky and Missouri had Confederate governments in exile and were represented in the Confederate Congress and Confederate battle flag. West Virginia was formed in 1863 after the western region broke away from Virginia, and fought off efforts of Virginia to recapture the region.

The popular definition of the "South" is more informal and is generally associated with those states that seceded during the Civil War to form the Confederate States of America. Those states share commonalities of history and culture that carry on to the present day.

Biologically, the South is a vast, diverse region, having numerous climatic zones, including temperate, sub-tropical, tropical, and arid—though the South is generally regarded as being hot and humid, with long summers and short mild winters, being significantly warmer than the regions to its north (and generally exhibiting the nation's highest heat indices). Many crops grow easily in its soils and can be grown without frost for at least six months of the year.

Some parts of the South, particularly the Southeast, have landscapes characterized by the presence of live oaks, magnolia trees, yellow jessamine vines, Spanish moss, cabbage palms and flowering dogwoods. Another common environment is the bayous and swampland of the Gulf Coast, especially in Louisiana and Texas.

The South is a victim of kudzu, an invasive fast-growing vine which covers large amounts of land and kills indigenous plant life. Kudzu is a particularly big problem in the piedmont regions of Mississippi, Alabama, and Georgia.

History

The first well-dated evidence of human occupation in the south United States occurs around 9500 BC with the appearance of the earliest documented Americans, who are now referred to as Paleo-Indians. Paleoindians were hunter-gathers that roamed in bands and frequently hunted megafauna. Several cultural stages, such as Archaic (ca. 8000-1000 BC) and the Woodland (ca. 1000 BC-AD 1000), preceded what the Europeans found at the end of the 15th century — the Mississippian culture.

The Mississippian culture was a complex, mound-building Native American culture that flourished in what is now the southeastern United States from approximately 800 AD to 1500 AD. Natives had elaborate and lengthy trading routes connecting their main residential and ceremonial centres extending through the river valleys and from the East Coast to the Great Lakes. Some noted explorers who encountered and described the Mississippian culture, by then in decline, included Panfilo de Narvaez (1528), Hernando de Soto (1540), and Pierre Le Moyne d'Iberville (1699).

Native American descendants of the mound-builders include Alabama, Apalachee, Caddo, Cherokee, Chickasaw, Choctaw, Creek, Guale, Hitchiti, Houma, and Seminole peoples, many of whom still reside in the South.

European Colonization

The predominant culture of the South was rooted in the settlement of the region by British colonists. In the seventeenth century, most voluntary immigrants were of English origins who settled chiefly along the coastal regions of the Eastern seaboard. The majority of early British settlers were indentured servants, who gained freedom after enough work to pay off their passage. The wealthier men who paid their way received land grants known as headrights, to encourage settlement.

The French and Spanish established colonies in Florida, Louisiana, and Texas. The Spanish colonized Florida in the 16th century, with their communities reaching a peak in the late 17th century. In the British and French colonies, most immigrants arrived after 1700. They cleared land, built houses and outbuildings, and worked on the large plantations that dominated export agriculture. Many were involved in the labour-intensive cultivation of tobacco, the first cash crop of Virginia. With a decrease in the number of British willing to go to the colonies in the eighteenth century, planters began importing more enslaved Africans, who became the predominant labour force on the plantations. Tobacco exhausted the soil quickly, requiring new fields to be cleared on a regular basis. Old fields were used as pasture and for crops such as corn and wheat, or allowed to grow into woodlots.

Also in the seventeenth century, colonists started importing African labourers. In the coastal and other settlements, early workers lived closely together in a multiracial society. Europeans married and made unions with Africans and Native Americans. The colonies gradually passed laws that hardened early conditions of indenture into lifelong racial slavery attached to African descent. Africans contributed to the economy of rice and indigo cultivation with their skilled knowledge, technology and labour, as well as to all the commodity crops; and to every aspect of culture (food, music, stories and religion).

Rice cultivation in South Carolina became another major commodity crop. Some historians have argued that slaves from the lowlands of western Africa, where rice was a basic crop, provided key skills, knowledge and technology for irrigation and construction of earthworks to support rice cultivation. The early methods and tools used in South Carolina were congruent with those in Africa. British immigrants would have had little or no familiarity with the complex process of growing rice in fields flooded by irrigation works. Africans were instrumental in the development of major earthworks for cultivating these commodities, as well as in the knowledge of technology and techniques for processing. The earthworks included extensive, elaborate systems of dams and irrigation for rice.

In the mid-to late-18th century, large groups of Ulster-Scots (later called the Scotch-Irish) and people from the Anglo-Scottish border region immigrated and settled in the back country of Appalachia and the Piedmont. They were the largest group of non-English immigrants from the British Isles before the American Revolution. In a census taken in 2000 of Americans and their self-reported ancestries, areas where people reported

'American' ancestry were the places where, historically, many Scotch-Irish, Scottish, and English Borderer Protestants settled in America, and especially the Appalachian region. The population with some Scots and Scotch-Irish ancestry may number 47 million, as most people have multiple heritages, some of which they may not know.

The early colonists, especially the Scotch-Irish in the back-country, engaged in warfare, trade, and cultural exchanges. Those living in the backcountry were more likely to join with Creek Indians, Cherokee, and Choctaws and other regional native groups.

The oldest university in the South, The College of William & Mary, was founded in 1693 in Virginia; it pioneered in the teaching of political economy and educated future U.S. Presidents Jefferson, Monroe and Tyler, all from Virginia. Indeed, the entire region dominated politics in the First Party System era: for example, four of the first five Presidents—Washington, Jefferson, Madison, and Monroe — were from Virginia. The two oldest public universities are also in the South: the University of North Carolina (1795) and the University of Georgia (1785).

American Revolution

The American Revolution provided a shock to slavery in the South. Tens of thousands of slaves took advantage of wartime disruption to find their own freedom, catalyzed by the British governor Dunmore of Virginia's promise of freedom for service. Many others simply escaped. Estimates are that five thousand slaves escaped from the Chesapeake Bay area, and thirteen thousand from South Carolina reached the British. "The extent of the loss to the slave owners in the lower South is indicated by the sharp decline between 1770 and 1790 in the proportion of population made up of black people (almost all of whom were slaves): from 60.5 percent to 43.8 percent in South Carolina and from 45.2 percent to 36.1 percent in Georgia."

In addition, some slaveholders were inspired to free their slaves after the Revolution. They were moved by the principles of the Revolution, and Quaker and Methodist preachers worked to encourage slaveholders to free their slaves. Planters often freed slaves by their wills. In the upper South, more than

10 percent of all blacks were free by 1810, a significant expansion from pre-war proportions of less than 1 percent free.

Antebellum Years

Cotton became dominant in the lower South after 1800. After the invention of the cotton gin, short staple cotton could be grown more widely. This led to an explosion of cotton cultivation, especially in the frontier uplands of Georgia, Alabama and other parts of the Deep South, as well as river front areas of the Mississippi Delta. Migrants poured into those areas in the early decades of the 19th century, when county population figures rose and fell as swells of people kept moving west. The expansion of cotton cultivation required more slave labour, and the institution became even more deeply an integral part of the South's economy.

With the opening up of frontier lands after the government forced most Native Americans to move west of the Mississippi, there was a major migration of both whites and blacks to those territories. From the 1820s through the 1850s, more than one million enslaved African Americans were transported to the Deep South in forced migration, two-thirds of them by slave traders and the others by masters who moved there. Planters in the Upper South sold slaves excess to their needs as they shifted from tobacco to mixed agriculture. Many enslaved families were broken up, as planters preferred mostly strong males for field work.

Two major political issues that festered in the first half of the 19th century caused political alignment along sectional lines, strengthened the identities of North and South as distinct regions with certain strongly opposed interests, and fed the arguments over states' rights that culminated in secession and the Civil War. One of these issues concerned the protective tariffs enacted to assist the growth of the manufacturing sector, primarily in the North. In 1832, in resistance to federal legislation increasing tariffs, South Carolina passed an ordinance of nullification, a procedure in which a state would in effect repeal a Federal law. Soon a naval flotilla was sent to Charleston harbour, and the threat of landing ground troops was used to compel the collection of tariffs. A compromise was reached by which the tariffs would be gradually reduced, but the underlying argument over states' rights continued to escalate in the following decades.

The second issue concerned slavery, primarily the question of whether slavery would be permitted in newly admitted states. The issue was initially finished by political compromises designed to balance the number of "free" and "slave" states. The issue resurfaced in more virulent form, however, around the time of the Mexican–American War, which raised the stakes by adding new territories primarily on the Southern side of the imaginary geographic divide. Congress opposed allowing slavery in these territories.

Before the Civil War, the number of immigrants arriving at Southern ports began to increase, although the North continued to receive the most immigrants. Numerous Irish immigrants flooded New Orleans, so much so that one of the sections of the city became known as the Irish Channel. Germans also went to New Orleans and its environs, resulting in a large area north of the city (along the Mississippi) becoming known as the German Coast; however, still greater numbers immigrated to Texas (especially after 1848), where many bought land and were farmers. Many more German immigrants arrived in Texas after the Civil War, where they created the brewing industry in Houston and elsewhere, became grocers in numerous cities, and also established wide areas of farming.

Tennessee was the last state to secede from the union, and it was the first to rejoin after the war.

Civil War

By 1856, the South was losing political power to the more populated North and was locked in a series of constitutional and political battles with the North regarding states' rights and the status of slavery in the territories. President James K. Polk imposed a low-tariff regime on the country (Walker Tariff of 1846), which angered Pennsylvania industrialists, and blocked proposed federal funding of national roads and port improvements. Once the North came to power in 1861, many Southerners felt it was time to secede from the union.

Seven cotton states decided on secession after the election of Abraham Lincoln in 1860 (often known as the pre-Sumter

Seven). They formed the Confederate States of America. In early 1861, they were joined by four more states immediately following the firing on Fort Sumter (splinter governments from two more states, Missouri and Kentucky, had governments-in-exile that were represented in Congress.) The United States government refused to recognize the seceding states and operated Fort Sumter, in Charleston harbour. Unable to claim true independence if a foreign fort dominated its most visible harbour, the Confederacy attacked and captured Ft. Sumter in April 1861. Immediately there was an unstoppable cry for war, North and South; only Kentucky tried to stay out of it.

In the four years of war 1861-65 the South was the primary battleground, with all but two of the major battles taking place on Southern soil. Union forces relentlessly squeezed the new nation, capturing the border states in 1861, the Tennessee River and New Orleans in 1863, and the Mississippi River in 1863. In the East, however, the rebel army under Robert E. Lee held off attack after attack in its defence of the national capital at Richmond. But when Lee tried to move north, he was repulsed (and nearly captured) at Antietam (1862) and Gettysburg (1863).

The Confederacy had the resources for a short war, but was unable to finance or supply a longer war. It reversed the traditional low-tariff policy of the South by imposing a new 15% tax on all imports from the North. The Union blockade stopped most commerce from entering the South, and smugglers avoided the tax, so the Confederate tariff produced too little revenue top finance the war. Inflated currency was the solution, but that created distrust of the Richmond government. Because of low investment in railroads, the Southern transportation system depended primarily on river and coastal traffic by boat; both were shut down by the Union Navy. The small railroad system virtually collapsed, so that by 1864 internal travel was so difficult that the Confederate economy was crippled.

The Confederate cause was hopeless by the time Atlanta fell and William T. Sherman marched through Georgia in 1864, but the rebels fought on, refusing to give up their independence. The South suffered much more than the North overall, as the Union strategy of attrition warfare meant that Lee could not replace his casualties, and the total war waged by Sherman and many other Union armies devastated the infrastructure and caused widespread poverty and distress. The Confederacy suffered military losses of 95,000 men killed in action and 165,000 who died of disease, for a total of 260,000, out of a total white Southern population at the time of around 5.5 million. Based on 1860 census figures, 8% of all white males aged 13 to 43 died in the war, including 6% in the North and about 18% in the South. Northern military casualties exceeded Southern casualties in absolute numbers, but were two-thirds smaller in terms of proportion of the population affected.

Reconstruction and Jim Crow

After the Civil War the South was devastated in terms of population, infrastructure and economy. Because of states' reluctance to grant voting rights to freedmen, Congress instituted Reconstruction governments. It established military districts and governors to rule over the South until new governments could be established. Many white Southerners who had actively supported the Confederacy were temporarily disfranchised. Rebuilding was difficult as people grappled with the effects of a new labour economy of a free market in the midst of a widespread agricultural depression. In addition, what limited infrastructure the South had was mostly destroyed by the war. At the same time, the North was rapidly industrializing.

There were thousands of people on the move, as African Americans tried to reunite families separated by slaves sales, and sometimes migrated for better opportunities in towns or other states. Other freed people moved from plantation areas to cities or towns for a chance to get different jobs and out from under white control. At the same time, whites returned from refuges to reclaim plantations or town dwellings. In some areas, many whites returned to the land to farm for a while. Some freed people left the South altogether for states such as Ohio and Indiana, and later, Kansas. Thousands of others joined the migration to new opportunities in the Mississippi and Arkansas Delta bottomlands and Texas.

With passage of the 13th Amendment to the Constitution of the United States (which outlawed slavery), the 14th Amendment (which granted full U.S. Citizenship to African Americans) and the 15th amendment (which extended the right to vote to African American males), African Americans in the South were made free citizens and were given the right to vote. Under Federal protection, white and black Republicans formed constitutional conventions and state governments. Among their accomplishments was creating the first public education systems in Southern states, and providing for welfare through orphanages, hospitals and similar institutions.

Northerners came south to participate in politics and business. Some were representatives of the Freedmen's Bureau and other agencies of Reconstruction; some were humanitarians with the intent to help black people. Also, as is often the case in volatile environments, some were adventurers who hoped to benefit themselves by questionable methods. They were all condemned with the pejorative term of carpetbagger. Some Southerners also took advantage of the disrupted environment and made money off various schemes, including bonds and financing for railroads.

Secret vigilante organizations such as the Ku Klux Klan—an organization sworn to perpetuate white supremacy—had arisen quickly after the war's end and used lynching, physical attacks, house burnings and other forms of intimidation to keep African Americans from exercising their political rights. Although the Klan was defeated by prosecution by the Federal government in the early 1870s, other groups persisted. By the mid-to late-1870s, elite white Southerners created increasing resistance to the altered social structure. Paramilitary organizations such as the White League in Louisiana (1874), the Red Shirts in Mississippi (1875) and rifle clubs, all "White Line" organizations, used organized violence against Republicans, blacks and whites, to turn Republicans out of office, repress and bar black voting, and restore Democrats to power. In 1876 white Democrats regained power in most of the state legislatures. They began to pass laws designed to strip African Americans and poor whites from the voter registration rolls. The success of late-19th century interracial coalitions in several states inspired a reaction among some white Democrats, who worked harder to prevent both groups from voting.

Despite discrimination, many blacks became property owners in areas that were still developing. For instance, 90% of the Mississippi's bottomlands were still frontier and undeveloped after the war. By the end of the century, two-thirds of the farmers in Mississippi's Delta bottomlands were black. They had cleared the land themselves and often made money in early years by selling off timber. Tens of thousands of migrants went to the Delta, both to work as labourers to clear timber for lumber companies, and many to develop their own farms.

Nearly all Southerners, black and white, suffered as a result of the Civil War. Within a few years cotton production and harvest was back to prewar levels, but low prices through much of the 19th century hampered recovery. They encouraged immigration by Chinese and Italian labourers into the Mississippi Delta. While the first Chinese entered as indentured labourers from Cuba, the majority came in the early-20th century. Neither group stayed long at rural farm labour. The Chinese became merchants and established stores in small towns throughout the Delta, establishing a place between white and black.

Migrations continued in the late-19th and early 20th-centuries among both blacks and whites. In the last two decades oth the 19th century about 141,000 blacks left the South, and more after 1900, totalling a loss of 537,000. After that the movement increased in what became known as the Great Migration from 1910–1940, and the Second Great Migration through 1970. Ven more whites left the South, some going to California for opportunities and others heading to Northern industrial cities after 1900. Between 1880 and 1910, the loss of whites totalled 1,243,000. Five million more left between 1940 and 1970.

From 1890 to 1908, 10 of the 11 states passed disfranchising constitutions or amendments which had provisions for voter registration, such as poll taxes, residency requirements and literacy tests, which were hard for many poor to meet. Most African Americans, Mexican Americans and tens of thousands of poor whites were disfranchised, losing the vote for decades. In some states grandfather clauses were temporarily used to exempt white illiterates from literacy tests. The numbers of voters dropped drastically throughout the South as a result. This can be seen on the feature "Turnout in Presidential and Midterm Elections" at the University of Texas Politics: Barriers to Voting. Alabama, which had established universal white suffrage in 1819 when it became a state, also substantially reduced voting by poor whites. Legislatures passed Jim Crow laws to segregate public facilities and services, including transportation.

While African Americans, poor whites and civil rights groups started litigation against such provisions in the early-20th century, for decades Supreme Court decisions overturning such provisions were rapidly followed by new state laws with new devices to restrict voting. Most blacks in the South could not vote until 1965, after passage of the Voting Rights Act and Federal enforcement to ensure people could register. Not until the late 1960s did all American citizens regain protected civil rights by passage of legislation following the leadership of the American Civil Rights Movement.

Late 19th and 20th Century—Industrialization and Great Migration

At the end of the 19th century, white Democrats in the South had created state constitutions that were hostile to industry and business development. Banking was limited, as was access to credit. States persisted in agricultural economies. As in Alabama, rural minorities held control in many state legislatures long after population had shifted to industrializing cities, and the legislators resisted business and modernizing interests. For instance, Alabama refused to redistrict from 1901 to 1972, long after major population and economic shifts to cities. For decades Birmingham generated the majority of revenue for the state, for instance, but received little back in services or infrastructure.

In the late 19th century, Texas rapidly expanded its railroad network, creating a network of cities connected on a radial plan and linked to the port of Galveston. It was the first state in which urban and economic development proceeded independently of rivers, the primary transportation network of the past. A reflection of increasing industry were strikes and labour unrest: "in 1885 Texas ranked ninth among forty states in number of workers involved in strikes (4,000); for the six-year period it ranked fifteenth. Seventy-five of the 100 strikes, chiefly interstate strikes of telegraphers and railway workers, occurred in the year 1886." In 1890 Dallas was the largest city in Texas. By 1900 it had a population of more than 42,000, which more than doubled to over 92,000 a decade later. Dallas was the harness making capital of the world and centre of other manufacturing. As an example of its ambitions, in 1907 Dallas built the Praetorian Building, 15 stories tall and the first skyscraper west of the Mississippi. Others soon followed. Texas was transformed by a railroad network linking five important cities, among them Houston with its nearby port at Galveston, Dallas, Fort Worth, San Antonio, and El Paso. Each exceeded 50,000 in population by 1920, with the major cities having three times that population.

Business interests were ignored by the Bourbon class. Nonetheless, major new industries started developing in cities such as Atlanta, GA; Birmingham, AL; and Dallas, Fort Worth and Houston, Texas. Growth began occurring at a geometric rate. Birmingham became a major steel producer and mining town, with major population growth in the early decades of the 20th century.

The first major oil well in the South was drilled at Spindletop near Beaumont, Texas, on the morning of January 10, 1901. Other oil fields were later discovered nearby in Arkansas, Oklahoma, and under the Gulf of Mexico. The resulting "Oil Boom" permanently transformed the economy of the West/South Central states and led to the most significant economic expansion after the Civil War.

In the early 20th century, invasion of the boll weevil devastated cotton crops in states of the South. This was an additional catalyst to African Americans' decisions to leave the South. From 1910 to 1940, and then from the 1940s to 1970, more than 6.5 million African Americans left the South in the Great Migration to northern and midwestern cities, making multiple acts of resistance against persistent lynching and violence, segregation, poor education, and inability to vote. Their movements transformed many cities, creating new cultures and music in the North. Many African Americans, like other groups, became industrial workers; others started their own businesses within the communities. Southern whites also migrated to industrial cities, especially Chicago and Detroit, where they took jobs in the booming new auto industry.

Later the southern economy was dealt additional blows by the Great Depression and the Dust Bowl. After the Wall Street Crash of 1929, the economy suffered significant reversals and millions were left unemployed.

Beginning in 1934 and lasting until 1939, an ecological disaster of severe wind and drought caused an exodus from Texas and Arkansas, the Oklahoma Panhandle region and the surrounding plains, in which over 500,000 Americans were homeless, hungry and jobless. Thousands left the region forever to seek economic opportunities along the West Coast.

President Franklin D. Roosevelt noted the South as the "number one priority" in terms of need of assistance during the Great Depression. His administration created programs such as the Tennessee Valley Authority in 1933 to provide rural electrification and stimulate development.

Locked into low productivity agriculture, the region's growth was slowed by limited industrial development, low levels of entrepreneurship, and the lack of capital investment.

World War II marked a time of change in the South as new industries and military bases were developed by the Federal government, providing badly needed capital and infrastructure in many regions. People from all parts of the US came to the South for military training and work in the region's many bases and new industries. Farming shifted from cotton and tobacco to include soybeans, corn, and other foods.

This growth increased in the 1960s and greatly accelerated into the 1980s and 1990s. Large urban areas with over 4 million people rose in Texas, Georgia, and Florida. Rapid expansion in industries such as autos, telecommunications, textiles, technology, banking, and aviation gave some states in the South an industrial strength to rival large states elsewhere in the country.

By the 2000 census, The South (along with the West) was leading the nation in population growth. However, with this growth has come long commute times and serious air pollution problems in cities such as Dallas, Houston, Atlanta, Miami, Austin, Charlotte, and others which have relied on sprawling development and highway networks.

Growth and Poverty

The South's early cash crops of tobacco, indigo and rice created enormous wealth for many planters in the coastal areas. While city development was limited, Richmond, Charleston, Savannah and others developed a sophisticated society. The wealthiest planters sent their sons to college in England and later to the best schools in the South, and sometimes the North. They imported furniture and furnishings from Europe, as well as employing the best colonial craftsmen. For many it was a mostly rural society, but by the 19th century, some families moved back and forth between plantations and town houses. The planter class controlled the state legislatures and kept taxes low. Their wealth went mostly for private purposes. They invested in no system of public education and little infrastructure.

In the antebellum years, by 1840 New Orleans was the wealthiest city in the country and the third largest in population, based on the growth of international trade associated with products being shipped to and from the interior of the country down the Mississippi River. It had the largest slave market in the country, as traders brought slaves to New Orleans by ship and overland to sell to planters across the Deep South. The city was a cosmopolitan port with a variety of jobs that attracted more immigrants than did other areas of the South. Because of lack of investment, construction of railroads to span the region lagged behind that in the North. People relied most heavily on river traffic for getting their crops to market and for transportation.

In Mississippi before the war, for instance, most plantations were developed along the Mississippi and other navigable rivers. The bottomlands were not developed until after the war, when the chance to buy land attracted tens of thousands of migrants, both black and white. By the end of the century, two-thirds of farm owners in the Delta bottomlands were black. The long agricultural depression meant that many had to take on too much debt—together with disfranchisement and lack of access to credit, by 1910 many had lost their property and by 1920, most blacks in the Delta were sharecroppers or landless workers. More than two generations of free African Americans had lost their stake in property.

After the Civil War, nearly the entire economic infrastructure of the region was in ruins. As agriculture had been the foundation of the Southern economy, disruption of slavery by the Civil War meant that planters had to learn to deal with free labour, a challenge as freedmen wanted most to take care of their own crops and land. Additionally, since there were few industrial businesses located in the south, there were not many other possible sources of income. Textile mills in the Piedmont of Georgia rebuilt rapidly, but it was not until the 20th century that the region dominated the industry. Some areas rapidly rebuilt—Atlanta, for example—through railroads.

After World War II, with the development of the Interstate Highway System, household air conditioning and later, passage of civil rights bills, the South was successful in attracting industry and business from other parts of the country. Industry from the Rust Belt region of the Northeast and the Great Lakes moved into the region because of lower labour costs and less unionization. Poverty rates and unemployment declined as a result of new job growth. Federal programs such as the Appalachian Regional Commission also contributed to economic growth.

While the Southern United States has advanced considerably since World War II, significant poverty still persists in the more isolated and rural areas. Areas like the Black Belt, the eastern Kentucky and southern West Virginia areas in Appalachia, the Mexican border area along the Rio Grande in Texas, and the Deltas of Mississippi and Arkansas suffer the most poverty in the South today.

Economy

In the last two generations, the South has changed dramatically. In recent decades it has seen a boom in its service economy, manufacturing base, high technology industries, and the financial sector. Examples of this include the surge in tourism in Florida and along the Gulf Coast; numerous new automobile production plants such as Mercedes-Benz in Tuscaloosa, Alabama; Hyundai in Montgomery, Alabama; the BMW production plant in Spartanburg, South Carolina; the GM manufacturing plant in Spring Hill, Tennessee; and the Nissan North American headquarters in Franklin, Tennessee;

the two largest research parks in the country: Research Triangle Park in North Carolina (the world's largest) and the Cummings Research Park in Huntsville, Alabama (the world's fourth largest); and the corporate headquarters of major banking corporations Bank of America and Wachovia in Charlotte; Regions Financial Corporation, AmSouth Bancorporation, and BBVA Compass in Birmingham; Sun Trust Banks and the district headquarters of the Federal Reserve Bank of Atlanta; and BB&T in Winston-Salem; and several Atlanta-based corporate headquarters and cable television networks, such as CNN, TBS, TNT, Turner South, Cartoon Network, and The Weather Channel. This economic expansion has enabled parts of the South to boast of some of the lowest unemployment rates in the United States.

Culture

The predominant culture of the South has its origins with the settlement of the region by British colonists in the 17th century, large groups of Scots and Ulster-Scots (later called the Scotch-Irish) who settled in Appalachia and the Piedmont in the 18th century, and the many African slaves who were part of the Southern economy. African-American descendants of the slaves brought into the South comprise the United States' second-largest racial minority, accounting for 12.1 percent of the total population according to the 2000 census. Despite Jim Crow era outflow to the North the majority of the black population remains concentrated in the southern states, and have heavily contributed to the cultural blend (the charismatic brand of Christianity, foods, art, music) that characterize Southern culture today.

Politics

In the first decades after Reconstruction, when white Democrats regained power in the state legislatures, they began to make voter registration more complicated, to reduce black voting. With a combination of intimidation, fraud and violence by paramilitary groups, they turned Republicans out of office and suppressed black voting. From 1890 to 1908, ten of eleven states ratified new constitutions or amendments that effectively disfranchised most black voters and many poor white voters. This disfranchisement persisted for six decades into the

20th century, depriving blacks and poor whites of all political representation. Because they could not vote, they could not sit on juries. They had no one to represent their interests, resulting in state legislatures consistently under funding programs and services, such as schools, for blacks and poor whites.

As the Supreme Court began to find such disfranchisement provisions unconstitutional, southern legislatures quickly passed other measures to keep blacks disfranchised, even after suffrage was extended more widely to poor whites. Because white Democrats controlled all the seats apportioned to their states, they had outsize power in Congress and filibustered or defeated efforts by others to pass legislation against lynching, for example. The region became known as the Solid South. The Republicans controlled parts of the Appalachian Mountains and competed for power in the Border States. From the late 1870s to the 1960s, it was rare for a state or national Southern politician to be Republican.

Increasing support for civil rights legislation by the national Democratic Party beginning in the 1940s caused conservative Southern Democrats to take notice. Until the passage of the Civil Rights laws of the 1960s, conservative Southern Democrats ("Dixiecrats") argued that only they could defend the region from the onslaught of northern liberals and the civil rights movement. In response to the Brown v. Board of Education ruling of 1954, southern legislators developed the Southern Manifesto. It was issued in March 1956, by 101 southern congressmen (19 senators, 82 House members of which 99 were Southern Democrats and 2 were Republicans). It denounced the Brown decisions as a "clear abuse of judicial power [that] climaxes a trend in the federal judiciary undertaking to legislate in derogation of the authority of Congress and to encroach upon the reserved rights of the states and the people." The manifesto lauded "those states which have declared the intention to resist enforced integration by any lawful means." It was signed by all southern senators except Majority Leader Lyndon B. Johnson, and Tennessee senators Albert Gore, Sr. and Estes Kefauver. Virginia closed schools in Warren County, Prince Edward County, Charlottesville, and Norfolk rather than integrate, but no other state followed suit. Democratic governors Orval Faubus of Arkansas, Ross Barnett of Mississippi, Lester

Maddox of Georgia, and, especially, George Wallace of Alabama resisted integration and appealed to a blue-collar electorate.

The Democratic Party's national support of civil rights issues culminated when Democratic President Lyndon B. Johnson signed into law the Civil Rights Act of 1964 and the Voting Rights Act of 1965. Some Republicans began to develop their Southern strategy to attract conservative white Southerners. Southern Democrats took notice that 1964 Republican Presidential candidate Barry Goldwater had voted against the Civil Rights Act. In the presidential election of 1964, Goldwater's only electoral victories outside his home state of Arizona were in the states of the Deep South.

The transition to a Republican stronghold in the South took decades. First, the states started voting Republican in presidential elections, except for native sons Jimmy Carter in 1976, Bill Clinton in 1992 and 1996. Then the states began electing Republican senators and finally governors. Georgia was the last state to do so, with Sonny Perdue taking the governorship in 2002. In addition to the middle class and business base, Republicans cultivated the religious right and attracted strong majorities from the evangelical Christian vote, which had not been a distinct political demographic prior to 1980.

The region's resistance to giving African Americans basic citizens' rights of voting and integration in public places broke out in renewed violence and murders during the 1960s, and major resistance to desegregation extending into the 1970s.

Presidential History

The South has produced the first winning presidential candidates for all but two major political parties in the history of the United States. The following is a list of presidents who represent their party's first candidate to reach the country's highest office:

- Democratic-Republicans: Thomas Jefferson of Virginia, in 1800.
- Democrats: Andrew Jackson, of Tennessee, in 1828.
- Whigs: William Henry Harrison, of Virginia, in 1840, and Zachary Taylor, of Louisiana, in 1848.

The exceptions are the Federalist Party which claimed its first (and only) presidential victory with John Adams of Massachusetts in 1796, and the Republican Party whose first victory was Abraham Lincoln of Illinois in 1860.

The South produced most of the U.S. Presidents prior to the Civil War. Memories of the war made it impossible for a Southerner to become President unless he either moved North (like Woodrow Wilson) or was a vice president who moved up (like Lyndon B. Johnson). In 1976, Jimmy Carter defied this trend and became the first Southerner to break the pattern since Zachary Taylor in 1848. The South produced five of the last eight American Presidents: Lyndon Johnson (1963–69), Jimmy Carter (1977–81), George H. W. Bush (1989–93), Bill Clinton (1993–2001) and George W. Bush (2001–2009). Carter was from Georgia, Clinton from Arkansas, and the others from Texas.

Other Politicians and Political Movements

The South has produced numerous other well-known politicians and political movements.

In 1948, a group of Democratic congressmen, led by Governor Strom Thurmond of South Carolina, split from the Democrats in reaction to an anti-segregation speech given by Senator Hubert Humphrey of Minnesota. They founded the States Rights Democratic or Dixiecrat Party. During that year's Presidential election, the party ran Thurmond as its candidate, but he was unsuccessful.

In the 1968 Presidential election, Alabama Governor George C. Wallace ran for President on the American Independent Party ticket. Wallace ran a "law and order" campaign similar to that of Republican candidate, Richard Nixon. Nixon's Southern Strategy of gaining electoral votes downplayed race issues and focused on culturally conservative values, such as family issues, patriotism, and cultural issues that appealed to Southern Baptists.

In 1994, another Southern politician, Newt Gingrich, ushered in 12 years of GOP control of the House. Gingrich became Speaker of the United States House of Representatives in 1995, but was forced to resign. Tom DeLay was the most powerful Republican leader in Congress until he was indicted under criminal charges in 2005 and was forced to step aside by Republican rules. Apart from Bob Dole of Kansas (1985-96), the recent Republican Senate leaders have been Southerners: Howard Baker (1981-85) of Tennessee, Trent Lott (1996-2003) of Mississippi, Bill Frist (2003-2006) of Tennessee, and Mitch McConnell (2007-present) of Kentucky.

The Republicans dominated Southern presidential elections from 1980 to 2004, with a slippage in 2008 as Democratic candidate Barack Obama carried Florida, Virginia and North Carolina.

Race Relations

Native Americans

Native Americans, had lived in the south for nearly 12,000 years. They were defeated by settlers in a series of wars ending in the War of 1812 and the Seminole War, and most were removed west to Indian Territory (now Oklahoma).

Civil Rights

In response to this treatment, the South witnessed two major events in the lives of 20th century African Americans: the Great Migration and the American Civil Rights Movement.

The Great Migration began during World War I, hitting its high point during World War II. During this migration, blacks left the racism and lack of opportunities in the South and settled in northern cities like Chicago, Detroit, Cleveland, Milwaukee, St. Louis, Pittsburgh, Philadelphia, New York City, and Boston, where they found work in factories and other sectors of the economy. (Katzman, 1996) However, Chicago quickly became the most segregated city in the north. This migration produced a new sense of independence in the Black community and contributed to the vibrant black urban culture seen during the Harlem Renaissance.

The migration also empowered the growing Civil Rights Movement. While the movement existed in all parts of the United States, its focus was against disfranchisement and the Jim Crow laws in the South. Most of the major events in the movement occurred in the South, including the Montgomery Bus Boycott, the Mississippi Freedom Summer, the March on

Selma, Alabama, and the assassination of Martin Luther King, Jr.. In addition, some of the most important writings to come out of the movement were written in the South, such as King's "Letter from Birmingham Jail". Most of the civil rights landmarks can be found around the South. The Martin Luther King, Jr., National Historic Site in Atlanta includes a museum that chronicles the American Civil Rights Movement as well as Martin Luther King, Jr.'s boyhood home on Auburn Avenue. Additionally, Ebenezer Baptist Church is located in the Sweet Auburn district as is the King Centre, location of Martin Luther and Coretta Scott King's gravesites.

As a result of the Civil Rights Movement, Jim Crow laws across the South were dropped. A second migration appears to be underway, with African Americans from the North moving to the South in record numbers. While race relations are still a contentious issue in the South, the region surpasses the rest of the country in many areas of integration and racial equality. According to 2003 report by researchers at the University of Wisconsin-Milwaukee, Virginia Beach, Charlotte, Nashville-Davidson, and Jacksonville were the four most integrated of the nation's fifty largest cities, with Memphis at number six. Southern states tend to have a low disparity in incarceration rates between blacks and whites relative to the rest of the country.

The Middle West—The Breadbasket of the Nation

The Midwestern United States (in the U.S. Generally referred to as the Midwest) is one of the four geographic regions within the United States of America used by the United States Census Bureau in its reporting.

The region consists of twelve states in the north-central United States: Illinois, Indiana, Iowa, Kansas, Michigan, Minnesota, Missouri, Nebraska, North Dakota, South Dakota, Ohio and Wisconsin.. A 2006 Census Bureau estimate put the population at 66,217,736. Both the geographic centre of the contiguous U.S. and the population centre of the U.S. are in the Midwest. The United States Census Bureau divides this region into the East North Central States (essentially the Great Lakes States) and the West North Central States. Chicago is the largest city in the region, followed by Detroit, Indianapolis, and Columbus. The Chicago-Joliet-Naperville, IL-IN-WI MSA is the largest metropolitan statistical area, followed by the Detroit-Warren-Livonia, MI MSA, and the Minneapolis-St. Paul-Bloomington, MN-WI MSA. Sault Ste. Marie, Michigan is the oldest city in the region, having been founded by French missionaries and explorers in 1668.

The term Midwest has been in common use for over 100 years. A variant term, "Middle West", has been in use since the 19th century and remains relatively common. Another term sometimes applied to the same general region is "the heartland". Other designations for the region have fallen into disuse, such as the "Northwest" or "Old Northwest" (from "Northwest Territory") and "Mid-America". Since the book Middletown appeared in 1929, sociologists have often used Midwestern cities (and the Midwest generally) as "typical" of the entire nation. The region has a higher employment-to-population ratio (the percentage of employed people at least 16 years old) than the Northeast, the West, the South, or the Sun Belt states.

Definition

Traditional definitions of the Midwest include the Northwest Ordinance "Old Northwest" states and many states that were part of the Louisiana Purchase. The states of the Old Northwest are also known as "Great Lakes states". Many of the Louisiana Purchase states are also known as "Great Plains states".

The North Central Region is defined by the U.S. Census Bureau as these 12 states:

- Illinois: Old Northwest, Ohio River, and Great Lakes state.
- Indiana: Old Northwest, Ohio River, and Great Lakes state.
- Iowa: Louisiana Purchase.
- Kansas: Louisiana Purchase, Great Plains state.
- Michigan: Old Northwest and Great Lakes state.
- Minnesota: Old Northwest, Louisiana Purchase, Great Lakes state.
- Missouri: Louisiana Purchase, Border state.
- Nebraska: Louisiana Purchase, Great Plains state.

- North Dakota: Louisiana Purchase, Great Plains state.
- Ohio: Old Northwest (Historic Connecticut Western Reserve), Ohio River, and Great Lakes state. The southeastern part of the state is part of Northern Appalachia.
- South Dakota: Louisiana Purchase, Great Plains state.
- Wisconsin: Old Northwest and Great Lakes state.

Physical Geography

While these states are for the most part relatively flat, consisting either of plains or of rolling and small hills, there is a measure of geographical variation. In particular, the eastern Midwest near the foothills of the Appalachian Mountains; the Great Lakes Basin; the Ozark Mountains of southern Missouri; the rugged topography of Southern Indiana and far Southern Illinois; and the Driftless Area of southwest Wisconsin, southeast Minnesota, and northeast Iowa exhibit a high degree of topographical variety. Prairies cover most of the states west of the Mississippi River with the exception of taiga-clad northern Minnesota. Illinois lies within an area called the "prairie peninsula", an eastward extension of prairies that borders deciduous forests to the north, east, and south. Rainfall decreases from east to west, resulting in different types of prairies, with the tallgrass prairie in the wetter eastern region, mixed-grass prairie in the central Great Plains, and shortgrass prairie towards the rain shadow of the Rockies. Today, these three prairie types largely correspond to the corn/soybean area, the wheat belt, and the western rangelands, respectively. Although hardwood forests in the northern Midwest were clear-cut in the late 1800s, they were replaced by new growth. Ohio and Michigan's forests are still growing. The majority of the Midwest can now be categorized as urbanized areas or pastoral agricultural areas.

Exploration and Early Settlement

European settlement of the area began in the 17th century following French exploration of the region. The French established a network of fur trading posts and Jesuit missions along the Mississippi River system and the upper Great Lakes. French control over the area ended in 1763 with the conclusion of the French and Indian War. British colonists began to expand into the Ohio Country during the 1750s. The Royal Proclamation of 1763 temporarily restrained expansion west of the Appalachian Mountains, but did not stop it completely.

Early settlement began either via routes over the Appalachian Mountains, such as Braddock Road, or through the waterways of the Great Lakes. Fort Pitt (now Pittsburgh) at the source of the Ohio River was an early outpost of the overland routes. The first settlements in the Midwest via the waterways of the Great Lakes were centered around military forts and trading posts such as Green Bay, Sault Ste. Marie, and Detroit. The first inland settlements via the overland routes were in southern Ohio or northern Kentucky, on either side of the Ohio River, and early such pioneers included Daniel Boone and Spencer Records.

Following the American Revolutionary War, the rate of settlers coming from the eastern states increased rapidly. In the 1790s, American Revolutionary War veterans and settlers from the original states moved there in response to federal government land grants. Among the earliest pioneers to Ohio and the Midwest were the Ulster-Scots Presbyterians of Pennsylvania (often through Virginia) and the Dutch Reformed, Quaker, and Congregationalists of Connecticut.

The region's fertile soil made it possible for farmers to produce abundant harvests of cereal crops such as corn, oats, and, most importantly, wheat. The region soon became known as the nation's "breadbasket".

Development of Transportation

Two waterways have been important to the development of the Midwest. The first and foremost was the Ohio River, which flowed into the Mississippi River. Development of the region was halted until 1795 due to Spain's control of the southern part of the Mississippi and its refusal to allow the shipment of American crops down the river and into the Atlantic Ocean.

The Mississippi River inspired two classic books – Life on the Mississippi and Adventures of Huckleberry Finn – written by native Missourian Samuel Clemens, who used the pseudonym Mark Twain. His stories became staples of Midwestern lore.

Twain's hometown of Hannibal, Missouri is a tourist attraction offering a glimpse into the Midwest of his time.

The second waterway is the network of routes within the Great Lakes. The opening of the Erie Canal in 1825 completed an all-water shipping route, more direct than the Mississippi, to New York and the seaport of New York City.

Lakeport cities grew up to handle this new shipping route. During the Industrial Revolution, the lakes became a conduit for iron ore from the Mesabi Range of Minnesota to steel mills in the Mid-Atlantic States. The Saint Lawrence Seaway (1862, widened 1959) opened the Midwest to the Atlantic Ocean.

Inland canals in Ohio and Indiana constituted another important waterway, which connected with Great Lakes and Ohio River traffic. The commodities that the Midwest funnelled into the Erie Canal down the Ohio River contributed to the wealth of New York City, which overtook Boston and Philadelphia. New York State would proudly boast of the Midwest as its "inland empire"; thus, New York would become known as the Empire State.

Transportation

During the mid-19th century the region got its first railroads, and the railroad junction in Chicago grew to be the world's largest. Even today, a century after Henry Ford, six Class I railroads meet in Chicago.

In the period from 1890 to 1930 many Midwestern cities, towns, villages, and even farms were connected by inter urbans, or electrical streetcars. The Midwest had more inter urbans than any other region. In 1916, Ohio led all states with 2,798 miles (4502 km), Indiana followed with 1,825 miles (2936 km). These two states alone had almost a third of the country's interurban trackage. The nation's largest interurban junction was in Indianapolis. During the first decade after the turn of the century the city's 38% growth in population was attributed largely to the interurban.

Competition with a growing population of automobiles and buses travelling on paved highways led to a decline in the interurban and other railroad passenger business. Henry Ford and Charles Kettering, the inventor of the electrical starting motor and leaded gasoline, were both products of the Midwest, as were the Wright brothers.

19th Century Sectional Conflict

The Northwest Ordinance region, comprising the heart of the Midwest, was the first large region of the United States that prohibited slavery (the Northeastern United States emancipated slaves in the 1830s). The regional southern boundary was the Ohio River, the border of freedom and slavery in American history and literature. The Midwest, particularly Ohio, provided the primary routes for the "Underground Railroad", whereby Midwesterners assisted slaves to freedom from their crossing of the Ohio River through their departure on Lake Erie to Canada.

The region was shaped by the relative absence of slavery (except for Missouri), pioneer settlement, education in one-room free public schools, democratic notions brought by American Revolutionary War veterans, Protestant faiths and experimentation, and agricultural wealth transported on the Ohio River riverboats, flatboats, canal boats, and railroads.

Industrialization and Immigration

By the time of the American Civil War, European immigrants bypassed the East Coast of the United States to settle directly in the interior: German immigrants to Ohio, Wisconsin, Minnesota, Illinois, and eastern Missouri; Irish immigrants to port cities on the Great Lakes, especially Chicago; Danes, Swedes, and Norwegians to Iowa, Wisconsin, Minnesota, and the Dakotas; and Finns to Upper Michigan and northern/central Minnesota. Poles, Hungarians, and Jews settled in Midwestern cities.

The U.S. was predominantly rural at the time of the Civil War. The Midwest was no exception, dotted with small farms all across the region. The late nineteenth century saw industrialization, immigration, and urbanization that fed the Industrial Revolution, and the heart of industrial domination and innovation was in the Great Lakes states of the Midwest, which only began its slow decline by the late twentieth century.

In the 20th century, African American migration from the Southern United States into the Midwestern states changed

Chicago, St. Louis, Cleveland, Milwaukee, Kansas City, Cincinnati, Gary, Detroit, Minneapolis, and many other cities in the Midwest dramatically, as factories and schools enticed families by the thousands to new opportunities.

History of the Term Midwest

As this region lies mostly in the eastern half of the United States, the term "Midwest" can be misleading if one does not understand American history.

The term West was applied to the region in the early years of the country. In 1789, the Northwest Ordinance was enacted, creating the Northwest Territory, which was bounded by the Great Lakes and the Ohio and Mississippi Rivers. Because the Northwest Territory lay between the East Coast and the then-far-West, the states carved out of it were called the "Northwest". In the early 19th century, anything west of the Mississippi River was considered the West, and the Midwest was the region east of the Mississippi and west of the Appalachians. In time, some users began to include Minnesota, Iowa and Missouri in the Midwest. With the settlement of the western prairie, the new term Great Plains States was used for the row of states from North Dakota to Kansas. Later, these states also came to be considered Midwest by some.

The states of the "old Northwest" are now called the "East North Central States" by the United States Census Bureau and the "Great Lakes" region by some of its inhabitants, whereas the states just west of the Mississippi and the Great Plains states are called the "West North Central States" by the Census Bureau. Today people as far west as eastern Colorado, Wyoming, and Montana, and southward into Oklahoma sometimes identify themselves with the term Midwest. Some parts of the Midwest are still referred to as "Northwest" for historical reasons – for example, Minnesota-based Northwest Airlines and Northwestern University in Illinois – so the Northwest region of the country is called the "Pacific Northwest" to make a clear distinction.

Culture

Roman Catholicism is the largest religious denomination in the Midwest, varying between 19 and 29% of the state populations. Southern Baptists compose 15.42% of Missouri's population and a small percentage in other Midwestern states. Lutherans are prevalent in the Upper Midwest, especially in Minnesota.

Judaism and Islam are each practiced by 1% or less of the population, with higher concentrations in major urban areas, such as Chicago, Cincinnati, St. Louis, Minneapolis, Detroit, and Cleveland. Those with no religious affiliation make up 13–16% of the Midwest's population. Around 50% of the people in the Midwest regularly attend church.

The rural heritage of the land in the Midwest remains widely held, even if industrialization and suburbanization have overtaken the states in the original Northwest Territory.

Because of 20th century African American migration from the South, a large African-American urban population lives in most of the region's major cities, although the concentration is not generally as large as that of the Southern United States. The combination of industry and cultures, jazz, blues, and rock and roll led to an outpouring of musical creativity in the 20th century, including new music genres such as the Motown Sound and techno from Detroit and house & blues music from Chicago. Additionally, the electrified Chicago blues sound exemplifies the genre, as popularized by record labels Chess and Alligator and portrayed in such films as The Blues Brothers, Godfathers and Sons and Adventures in Babysitting. Rock and roll music was first identified as a new genre by a Cleveland radio disc jockey, and the Rock and Roll Hall of Fame is located in Cleveland.

Cultural Overlap with Neighbouring Regions

Differences in the definition of the Midwest mainly split between the Heartland and the Great Plains on one side, and the Great Lakes and the Rust Belt on the other. While some point to the small towns and agricultural communities in Kansas, Iowa, the Dakotas, and Nebraska of the Great Plains as representative of traditional Midwestern lifestyles and values, others assert that the declining Rust Belt cities of the Great Lakes – with their histories of 19th-and early-20th-century immigration, manufacturing base, and strong Catholic influence – are more representative of the Midwestern experience.

Certain areas of the traditionally defined Midwest are often cited as not being representative of the region, while other areas traditionally outside of the Midwest are often claimed to be part of the Midwest. These claims often embody historical, cultural, economic or demographic arguments for inclusion or exclusion. Perceptions of the proper classification of the Midwest also vary within the region, and tend toward exclusion rather than inclusion.

Two other regions, Appalachia and the Ozark Mountains, overlap geographically with the Midwest – Appalachia in Southern Ohio and the Ozarks in Southern Missouri. The Ohio River has long been the boundary between North and South and between the Midwest and the Upper South. All of the lower Midwestern states, including Missouri, have a major Southern component, but only Missouri was a slave state before the Civil War.

Western Pennsylvania, which contains the cities of Erie and Pittsburgh and the Western New York city of Buffalo, New York, shares history with the Midwest but overlaps with Appalachia and the Northeast as well.

Although Kentucky is considered Midwestern by some, it is categorized as Southern by the Census Bureau.

Political Trends

One of the two major political parties in the United States, the Republican Party, originated in Ripon, in east-central Wisconsin, in the 1850s. It included opposition to the spread of slavery into new states as one of its agendas.

Midwestern political caution is sometimes peppered with protest, especially in minority communities or those associated with agrarian, labour, or populist roots. This was especially true in the early 20th century, when Milwaukee was a hub of the Socialist movement in the United States, electing three Socialist mayors and the only Socialist Congressional representative (Victor L. Berger) during that time. The metropolis-strewn Great Lakes region tends to be the most liberal area of the Midwest, and liberal presence diminishes gradually as one moves south and west from that region into the less-populated rural areas. The Great Lakes region has spawned politicians such as the La Follette political family, labour leader and five-time Socialist Party of America presidential candidate Eugene V. Debs, and Communist Party leader Gus Hall. Minnesota has produced liberal national politicians Paul Wellstone, Walter Mondale, Eugene McCarthy, and Hubert Humphrey, and protest musician Bob Dylan.

The region is now home to many critical swing states that do not have strong allegiance to either the Democratic or Republican party. Upper Midwestern states, such as Illinois, Minnesota, Wisconsin, and Michigan have proven reliably Democratic. Normally a Republican stronghold, Indiana became a key state in the 2006 mid-term elections, picking up three House Seats to bring the total to five Democrats to four Republicans representing Indiana in the U.S. House. In 2008, Indiana voted for the Democratic presidential candidate for the first time in 44 years. Prior to the 2008 presidential election, Missouri had supported the winning candidate all but once since the beginning of the 20th century.

The state government of Illinois is currently dominated by the Democratic Party. Both Illinois senators are Democrats and a majority of the state's U.S. Representatives are also Democrats. Illinois voters have preferred the Democratic presidential candidate by a significant margin in the past five elections (1992, 1996, 2000, 2004, 2008). The same is true of Michigan and Wisconsin, which also currently have Democratic governors and two Democratic senators.

Iowa is considered by many analysts to be the most evenly divided state in the country, but has leaned Democratic for the past fifteen years or more. Iowa has a Democratic governor, a Democratic Senator, three Democratic Congressmen out of five, has voted for the Democratic presidential candidate in four out of the last five elections, (1992, 1996, 2000, 2008). As of the 2006 midterm elections, Iowa has a state legislature dominated by Democrats in both chambers.

Minnesota voters have chosen the Democratic candidate for president longer than any other state. Minnesota was the only U.S. State (along with Washington, D.C.) to vote for Walter Mondale over Ronald Reagan in 1984 (Minnesota is Mondale's home state). In Iowa and Minnesota, however, the recent Democratic pluralities have often been fairly narrow. Minnesota has elected and re-elected a Republican governor, as well as supported some of the strongest gun concealment laws in the nation.

In 2006, Democrats scored major gains across the region. In Iowa, Democrats gained control of the state legislature and held onto the governor's mansion, giving them one-party control of Iowa's government. Elsewhere, Democrats gained control of the Wisconsin Senate, the Michigan Legislature, and the Indiana House. Minnesota, thought to be trending Republican, saw the Democratic-Farmer-Labour Party (DFL) post double-digit gains in the Minnesota House and win all state-wide elections, save for the gubernatorial race.

Democrats also won all state-wide races in Ohio, and gained control of all Illinois statewide offices. On a federal level, Democrat Sherrod Brown defeated Republican incumbent Mike DeWine 56-44 for the U.S. Senate.

By contrast, the Great Plains states of North Dakota, South Dakota, Nebraska, and Kansas have been strongholds for the Republicans for many decades. These four states have gone for the Republican candidate in every presidential election since 1940, except for Lyndon B. Johnson's landslide over Barry Goldwater in 1964 and Barack Obama's capture of one electoral vote in Nebraska in 2008.

However, North Dakota's Congressional delegation has been all-Democratic since 1987, and South Dakota has had at least two Democratic members of Congress in every year since 1987. Nebraska has elected Democrats to the Senate and as Governor in recent years, but the state's House delegation has been all-Republican since 1995. Kansas has elected a majority of Democrats as governor since 1956 but has not elected a Democrat to the Senate since 1932.

Missouri is considered a "bellwether state". Only twice since 1904 has the Show-Me-State not voted for the winner in the presidential election, in 1956 and in 2008. Missouri's House delegation has generally been evenly divided between Democrats and Republicans, with the Democrats holding sway in the large cities at the opposite ends of the state, Kansas City and St. Louis, and the Republicans controlling the rest of the state. Missouri's Senate seats were mostly controlled by Democrats until the latter part of the 20th century, but the Republicans have held one or both Senate seats continuously since the 1976 elections.

Around the turn of the 20th century, the region spawned the Populist movement in the Plains states and later the Progressive movement, which consisted largely of farmers and merchants intent on making government less corrupt and more receptive to the will of the people.

The Republicans were unified anti-slavery politicians, whose later interests in invention, economic progress, women's rights and suffrage, freedman's rights, progressive taxation, wealth creation, election reforms, temperance, and prohibition eventually clashed with the Taft–Roosevelt split in 1912. Similarly, the Populist and Progressive parties-including Roosevelt's 1912 Progressive Party ("Bull Moose Party") and Robert M. La Follette, Sr.'s 1924 Progressive Party-developed intellectually from the economic and social progress claimed by the early Republican party.

The Protestant and Midwestern ideals of profit, thrift, work ethic, pioneer self-reliance, education, democratic rights, and religious tolerance influenced both parties, despite their eventual drift into opposition.

Some in the Midwest favour isolationism, a belief that America should not involve itself in foreign entanglements. This position gained much support from German-and Swedish-American communities and leaders like Robert M. La Follette, Sr., Robert A. Taft, and Colonel Robert McCormick, publisher of the Chicago Tribune.

Linguistic Characteristics

The accents of the region are generally distinct from those of the South and of the urban areas of the American Northeast. To a lesser degree, they are also distinct from the accent of the American West.

The accent characteristic of most of the Midwest is considered by many to be "standard" American English. This accent is preferred by many national radio and television broadcasters.

This may have started because many prominent broadcast personalities – such as Walter Cronkite, Harry Reasoner,

Johnny Carson, David Letterman, Tom Brokaw, John Madden, Rush Limbaugh, and Casey Kasem – came from this region and so created this perception. A November 1998 National Geographic article attributed the high number of telemarketing firms in Omaha to the "neutral accents" of the area's inhabitants. Currently, many Midwestern cities are undergoing the Northern cities vowel shift away from the standard pronunciation of vowels.

The dialect of Minnesota, western Wisconsin, much of North Dakota and Michigan's Upper Peninsula is referred to as the Upper Midwestern Dialect (or "Minnesotan"), and has Scandinavian and Canadian influences.

2

The Western States—Their Resources and Industries

The Western United States, commonly referred to as the American West or simply "the West," traditionally refers to the region comprising the westernmost states of the United States. Because the U.S. Expanded westward after its founding, the meaning of the West has evolved over time. Prior to about 1800, the crest of the Appalachian Mountains was seen as the western frontier. Since then, the frontier moved further west and the Mississippi River was referenced as the easternmost possible boundary of the West.

In the 21st century, the states which include the Rocky Mountains and the Great Basin to the West Coast are generally considered to comprise the American West.

Besides being a purely geographical designation, "The West" also has anthropological connotations. While this region has its own internal diversity, there is arguably an overall shared history, culture (music, cuisine), mind set or world view and closely interrelated dialects of English. However, certain subregions and states, such as Utah and Southern California, have certain things that distinguish them from the other parts of the West.

The "West" had played an important part in American history; the Old West is embedded in America's folklore.

In its most extensive definition, the western U.S. is the largest region, covering more than half the land area of the United States. It is also the most geographically diverse, incorporating geographic regions such as the Pacific Coast, the temperate rainforests of the Northwest, the Rocky Mountains, the Great Plains, most of the tall-grass prairie eastward to Western Wisconsin, Illinois, the western Ozark Plateau, the western portions of the southern forests, the Gulf Coast, and all of the desert areas located in the United States (the Mojave, Sonoran, Great Basin, and Chihuahua deserts).

The states from the Rockies westward have something of a dual nature of semiarid steppes and arid deserts in the lowlands and plateaus, and mountains and coniferous forests in the uplands and coastal regions.

The region encompasses some of the Louisiana Purchase, most of the land ceded by Britain in 1818, some of the land acquired when the Republic of Texas joined the U.S., all of the land ceded by Britain in 1846, all of the land ceded by Mexico in 1848, and all of the Gadsden Purchase.

Arizona and New Mexico are always considered to be in the Southwest while portions of California, Colorado, Nevada, Oklahoma, Texas, and Utah are sometimes considered part of the Southwest, while all or part of Idaho, Montana, Oregon, Washington, and Wyoming can be considered part of the Northwest, more narrowly part or all of those same states, with the exception of Wyoming and the eastern portions of Montana and Idaho, and the addition of the Canadian province of British Columbia comprise the Pacific Northwest.

The West can be divided into the Pacific States; Alaska, California, Hawaii, Oregon, and Washington, with the term West Coast usually restricted to just California, Oregon, and Washington, and the Mountain States, always Arizona, Colorado, Idaho, Montana, Nevada, New Mexico, Utah, and Wyoming. Alaska and Hawaii, being detached from the other western states, have few similarities with them, but are usually also classified as part of the West. Western Texas in the Chihuahuan Desert is also traditionally considered part of the Western U.S, though from a climatological perspective the West might be said to begin just west of Austin, TX where annual rainfall drops off significantly from what is typically experienced in the East, with a concurrent change in plant and animal species.

Some western states are grouped into regions with eastern states. Kansas, Nebraska, South Dakota and North Dakota are often included in the Midwest, which also includes states like Iowa, Illinois and Wisconsin. Texas, Arkansas, and Louisiana—and to a lesser extent, Oklahoma—are also considered part of the South.

It is rare for any state east of the Mississippi River to be considered part of the modern west. Historically, however, the Northwest Territory was an important early territory of the U.S., comprising the modern states of Ohio, Indiana, Illinois, Michigan and Wisconsin, as well as the northeastern part of Minnesota.

Demographics

According to the 2000 Census, the West's population was:

- 68.5% White
- 12.1% of Some other race,
- 7.9% Asian,
- 4.9% Black or African American,
- 4.3% Two or more races,
- 1.8% American Indian and Alaska Native, and
- 0.5% Native Hawaiian and Pacific Islander.
- 24.3% were Hispanic or Latino (of any race).

As defined by the United States Census Bureau, the Western region of the United States includes 13 states (with a total 2006 estimated population of 69,355,643) and is split into two smaller units, or divisions:

- The Mountain States: Montana, Wyoming, Colorado, New Mexico, Idaho, Utah, Arizona, and Nevada
- The Pacific States: Washington, Oregon, California, Alaska and Hawaii.

However, the United States Census Bureau uses only one definition of the West in its reporting system, which may not coincide with what may be historically or culturally considered the West. For example, in the 2000 Census, the Census Bureau included the state with the second largest Hispanic population, Texas, in the South, included the state with the second largest American Indian population, Oklahoma, also in the South, and included the Dakotas, with their large populations of Plains Indians, in with the Midwest. However, it should be noted that the western half of Oklahoma and far West Texas, are usually neither culturally, geographically or socioeconomically identified with the South.

Statistics from the 2000 United States Census, adjusted to include the second tier of States west of the Mississippi, show that, under that definition, the West would have a population of 91,457,662, including 1,611,447 Indians, or 1.8% of the total, and 22,377,288 Hispanics (the majority Mexican), or 24.5% of the total. Indians comprise 0.9% of all Americans, and Hispanics, 12.5%. Asians, important from the very beginning in the history of the West, totalled 5,161,446, or 5.6%, with most living in the Far West. African-Americans, totalled 5,929,968, or 6.5%—lower than the national proportion (12.8%). The highest concentrations (12%) of black residents in the West are found in Texas—which is also considered a Southern state—and in California.

The West is still one of the most sparsely settled areas in the United States with 49.5 inhabitants per square mile (19/km^2). Only Texas with 78.0 inhabitants/sq mi. (30/km^2), Washington with 86.0 inhabitants/sq mi. (33/km^2), and California with 213.4 inhabitants/sq mi. (82/km^2) exceed the national average of 77.98 inhabitants/sq mi. (30/km^2).

The entire Western region has also been strongly influenced by European, Native and Hispanic culture; it contains the largest number of minorities in the U.S. and encompasses the only four American states where all racial groups including Caucasians are a minority (California, Hawaii, New Mexico, and Texas). While most of the studies of racial dynamics in America such as riots in Los Angeles have been written about European and African Americans, in many cities in the West and California, European and African Americans together are less than half the population because of the preference for the region by Hispanics and Asians. African and European Americans, however, continue to wield a stronger political influence because of the lower rates of citizenship and voting among Asians and Hispanics.

Because the tide of development had not yet reached most of the West when conservation became a national issue, agencies of the federal government own and manage vast areas of land. (The most important among these are the National Park Service and the Bureau of Land Management within the Interior Department, and the U. S. Forest Service within the Agriculture Department.) National parks are reserved for recreational activities such as fishing, camping, hiking, and boating, but other government lands also allow commercial activities like ranching, logging and mining. In recent years, some local residents who earn their livelihoods on federal land have come into conflict with the land's managers, who are required to keep land use within environmentally acceptable limits.

The largest city in the region is Los Angeles, located on the West Coast. Other West Coast cities include San Diego, San Jose, San Francisco, San Bernardino, Sacramento, Seattle, and Portland. Prominent cities in the Mountain States include Denver, Colorado Springs, Phoenix, Tucson, Albuquerque, Las Vegas, and Salt Lake City.

Natural Geography

Along the Pacific Ocean coast lie the Coast Ranges, which, while not approaching the scale of the Rockies, are formidable nevertheless. They collect a large part of the airborne moisture moving in from the ocean. Even in the relatively arid climate of Central California, the Coast Ranges squeeze enough water out of the clouds to support the growth of coast redwoods. East of the Coast Ranges lie several cultivated fertile valleys, notably the San Joaquin Valley of California and the Willamette Valley of Oregon.

Beyond the valleys lie the Sierra Nevada in the south and the Cascade Range in the north. These mountains are some of the highest in the United States. Mount Whitney, at 14,505 feet (4,421 m) the tallest peak in the contiguous 48 states, is in the Sierra Nevada. The Cascades are also volcanic. Mount Rainier, a volcano in Washington, is also well over 14,000 feet (4,300 m). Mount St. Helens, a volcano in the Cascades erupted explosively in 1980. A major volcanic eruption at Mount Mazama around 4860 BCE formed Crater Lake. These mountain ranges see heavy precipitation, capturing most of the moisture that remains after the Coast Ranges, and creating a rain shadow to the east forming vast stretches of arid land. These dry areas encompass much of Nevada, Utah and Arizona. The Mojave Desert and Sonoran Desert along with other deserts are found here.

Beyond the deserts lie the Rocky Mountains. In the north, they run almost immediately east of the Cascade Range, so that the desert region is only a few miles wide by the time one reaches the Canadian border. The Rockies are hundreds of miles wide, and run uninterrupted from New Mexico to Alaska. The tallest peaks of the Rockies, some of which are over 14,000 feet (4,250 meters approx.), are found in central Colorado.

The West has several long rivers that empty into the Pacific Ocean, while the eastern rivers run into the Gulf of Mexico. The Mississippi River forms the easternmost possible boundary for the West today. The Missouri River, a tributary of the Mississippi, flows from its headwaters in the Rocky Mountains eastward across the Great Plains, a vast grassy plateau, before sloping gradually down to the forests and hence to the Mississippi. The Colorado River snakes through the Mountain states, at one point forming the Grand Canyon. The Colorado is a major source of water in the Southwest and many dams, such as the Hoover Dam, form reservoirs along it. So much water is drawn for drinking water throughout the West and irrigation in California that in some years, water from the Colorado no longer reaches the Gulf of California. The Columbia River, the largest river in volume flowing into the Pacific Ocean from North America, and its tributary, the Snake River, water the Pacific Northwest. The Platte runs through Nebraska and was known for being a mile (2 km) wide but only a half-inch (1 cm) deep. The Rio Grande forms the border between Texas and Mexico before turning due north and splitting New Mexico in half.

According to the United States Coast Guard, "The Western Rivers System consists of the Mississippi, Ohio, Missouri, Illinois, Tennessee, Cumberland, Arkansas and White Rivers and their tributaries, and certain other rivers that flow towards the Gulf of Mexico."

Climate and Agriculture

As a very gross generalization, the climate of the West can be described as overall semiarid; however, parts of the West get extremely high amounts of rain and/or snow, and still other parts are true desert and get less than 10 inches of rain per year. Also, the climate of the West is quite unstable, and areas that are normally wet can be very dry for years and vice versa.

The seasonal temperatures vary greatly throughout the West. Low elevations on the West Coast have warm to very hot summers and get little to no snow. The Desert Southwest has very hot summers and mild winters. The Inland Northwest has a continental climate of warm to hot summers and cold to bitter cold winters.

Annual rainfall is greater in the eastern portions, gradually tapering off until reaching the Pacific Coast where it again increases. In fact, the greatest annual rainfall in the United States falls in the coastal regions of the Pacific Northwest. Drought is much more common in the West than the rest of the United States. The driest place recorded in the U.S. is Death Valley, California.

Violent thunderstorms occur east of the Rockies. Tornadoes occur every spring on the southern plains, with the most common and most destructive centered on Tornado Alley, which covers eastern portions of the West, (Texas to North Dakota), and all states in between and to the east.

Agriculture varies depending on rainfall, irrigation, soil, elevation, and temperature extremes. The arid regions generally support only livestock grazing, chiefly beef cattle. The wheat belt extends from Texas through the Dakotas, producing most of the wheat and soybeans in the U.S. and exporting more to the rest of the world. Irrigation in the Southwest allows the growing of great quantities of fruits, nuts, and vegetables as well as grain, hay, and flowers. Texas is a major cattle and sheep raising area, as well as the nation's largest producer of cotton. Washington is famous for its apples, and Idaho for its potatoes. California and Arizona are major producers of citrus crops, although growing metropolitan sprawl is absorbing much of this land.

Local state and Government officials started to understand, after several surveys made during the latter part of the nineteenth century, that only action by the federal government could provide water resources needed to support the development of the West. Starting in 1902, Congress passed a series of acts authorizing the establishment of the United States Bureau of Reclamation to oversee water development projects in seventeen western states.

During the first half of the 20th century, dams and irrigation projects provided water for rapid agricultural growth throughout the West and brought prosperity for several states, where agriculture had previously only been subsistence level. Following World War II, the West's cities experienced an economic and population boom. The population growth, mostly in the Southwest, has strained water and power resources, with water diverted from agricultural uses to major population centres, such as Las Vegas and Los Angeles.

Geology

Plains make up most of the eastern half of the West, underlain with sedimentary rock from the Upper Paleozoic, Mesozoic, and Cenozoic eras. The Rocky Mountains expose igneous and metamorphic rock from both the Precambrian and the Post Precambrian periods. The Inter-mountain States and Pacific Northwest have huge expanses of volcanic rock from the Cenozoic period. Salt flats and salt lakes reveal a time when the great inland seas covered much of what is now the West. The Pacific states are the most geologically active areas in the United States. Earthquakes cause major damage every few years in California. While the Pacific states are the most volcanically active areas, extinct volcanoes and lava flows are found throughout most of the western half of the West.

History and Culture

Facing both the Pacific Ocean and the Mexican border, the West has been shaped by a variety of ethnic groups. Hawaii is the only state in the union in which Asian Americans outnumber white American residents. Asians from many countries have settled in California and other coastal states in several waves of immigration since the 1800s, contributing to the Gold Rush, the building of the transcontinental railroad, agriculture, and more recently, high technology.

The southwestern border states—California, Arizona, New Mexico, and Texas—all have large Hispanic populations, and the many Spanish place names attest to their history as former

Spanish and Mexican territories. The West also contains much of the Native American population in the U.S., particularly in the large reservations in the mountain and desert states.

The largest concentrations for black Americans in the West can be found in Southern California, Las Vegas, and parts of Colorado and Arizona.

Alaska—the northernmost state in the Union—is a vast land of few people, many of them native, and of great stretches of wilderness, protected in national parks and wildlife refuges. Hawaii's location makes it a major gateway between the U.S. and Asia, as well as a centre for tourism.

In the Pacific Coast states, the wide areas filled with small towns, farms, and forests are supplemented by a few big port cities which have evolved into world centres for the media and technology industries. Now the second largest city in the nation, Los Angeles is best known as the home of the Hollywood film industry; the area around Los Angeles also was a major centre for the aerospace industry by World War II, though Boeing, located in Washington state would lead the aerospace industry. Fueled by the growth of Los Angeles—as well as the San Francisco Bay area, including Silicon Valley—California has become the most populous of all the states. Oregon and Washington have also seen rapid growth with the rise of Boeing and Microsoft along with agriculture and resource based industries. The desert and mountain states have relatively low population densities, and developed as ranching and mining areas which are only recently becoming urbanized. Most of them have highly individualistic cultures, and have worked to balance the interests of urban development, recreation, and the environment.

Culturally distinctive points include the large Mormon population in the Mormon Corridor, including southeastern Idaho, Utah, Northern Arizona and Nevada; the extravagant casino resort towns of Las Vegas and Reno, Nevada; and, of course, the many Native American tribal reservations.

American Old West

Major settlement of the western territories by migrants from the states in the east developed rapidly in the 1840s, largely through the Oregon Trail and the California Gold Rush of 1849; California experienced such a rapid growth in a few short months that it was admitted to statehood in 1850 without the normal transitory phase of becoming an official territory. The largest migration in American history occurred in the 1840s as the Latter Day Saints left the Midwest for the safety of the West. Both Omaha, Nebraska and St. Louis, Missouri laid claim to the title, "Gateway to the West" during this period. Omaha, home to the Union Pacific Railroad and the Mormon Trail, made its fortunes on outfitting settlers; St. Louis built itself upon the vast fur trade in the West before its settlement.

The 1850s were marked by political controversies which were part of the national issues leading to the Civil War, though California had been established as a non-slave state in the Compromise of 1850; California played little role in the war itself due to its geographic distance from major campaigns. In the aftermath of the Civil War, many former Confederate partisans migrated to California during the end of the Reconstruction period.

The history of the American West in the late nineteenth century and early twentieth century has acquired a cultural mythos in the literature and cinema of the United States. The image of the cowboy, the homesteader and westward expansion took real events and transmuted them into a myth of the west which has influenced American culture since at least the 1920s.

Writers as diverse as Bret Harte and Zane Grey celebrated or derided cowboy culture, while artists such as Frederic Remington created western art as a method of recording the expansion into the west. The American cinema, in particular, created the genre of the western movie, which, in many cases, use the West as a metaphor for the virtue of self-reliance and an American ethos.

The contrast between the romanticism of culture about the West and the actuality of the history of the westward expansion has been a theme of late Twentieth and early Twenty-First century scholarship about the West. Cowboy culture has become embedded in the American experience as a common cultural touchstone, and modern forms as diverse as country and western music and the works of artist Georgia O'Keeffe have celebrated the supposed sense of isolation and independence of spirit inspired by the unpopulated and relatively harsh climate of the region.

As a result of the various periods of rapid growth, many new residents were migrants who were seeking to make a new start after previous histories of either personal failure or hostilities developed in their previous communities. With these and other migrants who harbored more commercial goals in the opening country, the area developed a strong ethos of self-determinism and individual freedom, as communities were created whose residents shared no prior connection or common set of ideals and allegiances. The open land of the region allowed residents to live at a much greater distance from neighbours than had been possible in eastern cities, and an ethic of tolerance for the different values and goals of other residents developed. California's state constitutions (in both 1849 and 1879) were largely drafted by groups which sought a strong emphasis on individual property rights and personal freedom, arguably at the expense of ideals tending toward civic community.

The Twentieth Century

By 1890, the frontier was gone. The advent of the automobile enabled the average American to tour the West. Western businessmen promoted U.S. Route 66 as a means to bring tourism and industry to the West. In the 1950s, representatives from all the western states built the Cowboy Hall of Fame and Western Heritage Centre to showcase western culture and greet travellers from the East. During the latter half of the twentieth century, several transcontinental interstate highways crossed the West bringing more trade and tourists from the East. In the news, reports spoke of oil boom towns in Texas and Oklahoma rivalling the old mining camps for their lawlessness, of the Dust Bowl forcing children of the original homesteaders even further west. The movies replaced the dime novel as the chief entertainment source featuring western fiction.

Although there has been segregation, along with accusations of racial profiling and police brutality towards minorities due to issues such as illegal immigration and a racial shift in neighbourhood demographics, sometimes leading to racially based riots, the West has a continuing reputation for being open-minded and for being one of the most racially progressive areas in the United States

Politics

The region's distance from historical centres of power in the East, and the celebrated "frontier spirit" of its settlers offer two cliches for explaining the region's independent, heterogeneous politics. Historically, the West was the first region to see widespread women's suffrage. California birthed both the property rights and conservation movements, and spawned such phenomena as the Taxpayer Revolt and the Berkeley Free Speech Movement. It has also produced three U.S. presidents: Herbert Hoover, Richard Nixon and Ronald Reagan.

The prevalence of libertarian political thought, even if not labelled as such, can be widely observed. For example, the majority of Western states have legalized medicinal marijuana (all but Utah, and Wyoming) and some forms of gambling (except Utah); Oregon and Washington have legalized physician-assisted suicide; Utah has a long history of former polygamous territorial leaders; and most rural counties in Nevada allow licensed brothels. There is less resistance to the legal recognition of same-sex unions: California, Hawaii, Nevada, Oregon, and Washington recognize them, and only 28% of all western residents are against legal recognition (compared to the 48% in southern states).

The West Coast leans toward the Democratic Party. San Francisco's two main political parties are the Green Party and the Democratic Party. Seattle has historically been a centre of radical left-wing politics; the union Industrial Workers of the World is particularly active.

A former mayor of Salt Lake City, Rocky Anderson, supports same-sex marriage, and Denver's residents have legalized the private use of marijuana, but only for persons aged 21 and older (Colorado's age of majority is 18). Hawaii has come closest to adopting single payer healthcare financing in the U.S. Both the Democratic leaders of the U.S. Congress are from the region: Speaker Nancy Pelosi of California and Senate Majority Leader Harry Reid of Nevada.

Interior areas are more Republican, with Arizona, Idaho, Montana, Utah, and Wyoming being Republican strongholds, and Colorado, Nevada, and New Mexico being swing states. The state of Arizona has been won by the Republican presidential candidate in every election except one since 1948, while the states of Idaho, Utah, and Wyoming have been won by the Republican presidential candidate in every election since 1964.

As the fastest-growing demographic group, Latinos are hotly contested by both parties; immigration remains an important political issue for this group. Backlash against illegal immigration led to the passage of California Proposition 187 in 1994, a ballot initiative which would have denied many public services to illegal immigrants. Association of this proposal with the California Republicans, especially incumbent governor Pete Wilson, is credited with driving many Hispanic voters to the Democrats.

Conserving our resources—our obligation to coming generations.

Conservation in the United States can be traced back to the 19th century with the formation of the first ever National Park.

History

Philosophy of Early American Conservation Movement

During the nineteenth century, Americans developed a deep and abiding passion for nature. The early evolution of the conservation movement began through both public and private recognition of the relationship between man and nature often reflected in the great literary and artistic works of the nineteenth century. Artists, such as Albert Bierstadt, painted powerful landscapes of the American West during the mid nineteenth century, which were incredibly popular ages representative of the unique natural wonders of the American frontier. Likewise, in 1860, Frederic Edwin Church painted "Twilight in the Wilderness", which was an artistic masterpiece of the era that explored the growing importance of the American wilderness.

Many American writers also romanticized and focused upon nature as a subject matter. However, the most notable literary figure upon the early conservation movement proved to be

Henry David Thoreau. Throughout his work, Walden, Thoreau detailed his experiences at the natural setting of Walden Pond and his deep appreciation for nature. In one instance, he described a deep grief for a tree that was cut down. Thoreau went on to bemoan the lack of reverence for the natural world: "I would that our farmers when they cut down a forest felt some of that awe which the old Romans did when they came to thin, or let in the light to, a consecrated grove". As he states in Walden, Thoreau "was interested in the preservation" of nature. In 1860, Henry David Thoreau delivered a speech to the Middlesex Agricultural Society in Massachusetts; the speech, entitled "The Succession of Forest Trees", explored forest ecology and encouraged the agricultural community to plant trees. This speech became one of Thoreau's "most influential ecological contributions to conservationist thought".

The early conservation movement in the United States was also successful due to the hard work of John Muir. Muir was a former carriage worker who was nearly blinded by an accident at work. After almost losing his sight, Muir decided to see "America's natural wonders". Based upon his travels throughout Yosemite and the Sierra Nevada Mountains, Muir wrote a collection of articles for Century magazine, entitled "Studies in the Sierra". In 1892, John Muir joined forces with the editor of "Century" Magazine, Robert Underwood Johnson, to establish the Sierra Club, an organization designed to protect America's natural resources and public parks. Early Americans recognized the importance of natural resources and the necessity of wilderness preservation for sustained yield harvesting of natural resources. In essence, the preservation of wilderness and landscapes were recognized as critical for future generations and their continued subsistence in a healthy environment. The foundation of the conservation movement is grounded during this period between 1850 and 1920. Ultimately, historical trends and cultural mind-sets were united, which influenced ideas and policy towards the early history of the conservation movement in the United States.

Early American Conservation Movement

America had its own conservation movement in the 19th century, most often characterized by George Perkins Marsh, author of Man and Nature. The expedition into northwest

Wyoming in 1871 led by F. V. Hayden and accompanied by photographer William Henry Jackson provided the imagery needed to substantiate rumors about the grandeur of the Yellowstone region, and resulted in the creation of Yellowstone National Park, the world's first, in 1872. Travels by later U.S. President Theodore Roosevelt through the region around Yellowstone provided the impetus for the creation of the Yellowstone Timberland Reserve in 1891. The largest section of the reserve was later renamed Shoshone National Forest, and it is the oldest National Forest in the U.S. But it was not until 1898 when German forester Dr. Carl A. Schenck, on the Biltmore Estate, and Cornell University founded the first two forestry schools, both run by Germans. Bernard Fernow, founder of the forestry schools at Cornell and the University of Toronto, was originally from Prussia (Germany), and he honed his knowledge from Germans who pioneered forestry in India. He introduced Gifford Pinchot, the "father of American forestry", to Brandis and Ribbentrop in Europe. From these men, Pinchot learned the skills and legislative patterns he would later apply to America. Pinchot, in his memoir history Breaking New Ground, credited Brandis especially with helping to form America's conservation laws.

> *"Conservation means the greatest good to the greatest number for the longest time.—Pinchot"*

Pinchot wrote that the principles of conservation were:

1. Development: *"the use of the natural resources now existing on this continent for the benefit of the people who live here now. There may be just as much waste in neglecting the development and use of certain natural resources as there is in their destruction. ... The development of our natural resources and the fullest use of them for the present generation is the first duty of this generation."*
2. Conservation: *"...the prevention of waste in all other directions is a simple matter of good business. The first duty of the human race is to control the earth it lives upon."*
3. Protection of the public interests: *"The natural resources must be developed and preserved for the benefit of the many, and not merely for the profit of a few."*

In 1891, Congress passed the Forest Reserve Act, which allowed the President of the United States to set aside forest lands on public domain. A decade after the Forest Reserve Act, presidents Harrison, Cleveland, and McKinley had transferred approximately 50,000,000 acres (200,000 km^2) into the forest reserve system. However, President Theodore Roosevelt is credited with the institutionalization of the conservation movement in the United States.

For President Roosevelt, the conservation movement was not about the preservation of nature simply for nature itself. After his experiences travelling as an enthusiastic, zealous hunter, Roosevelt became convinced of "the need for measures to protect the game species from further destruction and eventual extinction". President Roosevelt recognized the necessity of carefully managing America's natural resources. According to Roosevelt, "We are prone to speak of the resources of this country as inexhaustible; this is not so". Nonetheless, Roosevelt believed that conservation of America's natural resources was for the successful management and continued sustain yield harvesting of these resources in the future for the benefit and enjoyment of the American people. Roosevelt took several major steps to further his conservation goals. In 1902, Roosevelt signed the National Reclamation Act, which allowed for the management and settlement of a large tract of barren land. Then, in 1905, President Roosevelt helped to create the United States Forest Service and then appointed respected forester, Gifford Pinchot, as the first head of the agency. By the end of his presidency, Theodore Roosevelt, in partnership with Gifford Pinchot, had successfully increased the number of national parks as well as added area to existing forest reserves.

Despite these advancements, the American conservation movement did have difficulties. In the early 1900s the conservation movement in America was split into two main groups: conservationists, like Pinchot and Roosevelt, who were utilitarian foresters and natural rights advocates who wanted to protect forests "for the greater good for the greatest length", and preservationists, such as John Muir, the founder of the Sierra Club. Important differences separated conservationists like Roosevelt and Pinchot from preservationists like Muir. As a preservationist, Muir envisioned the maintenance of pristine natural environments where any development was banned. Whereas conservationists wanted regulated use of forest lands for both public activities and commercial endeavours, preservationists wanted forest to be preserved for natural beauty, scientific study and recreation. The differences continue to the modern era, with sustainable harvest and multiple-use the major focus of the U.S. Forest Service and recreation emphasized by the National Park Service.

Modern American Conservation Movement

Ultimately, the modern conservation movement in the United States continues to strive for the delicate balance between the successful management of society's industrial progress while still preserving the integrity of the natural environment that sustains humanity. In a large part, today's conservation movement in the United States is a joint effort of individuals, grassroots organizations, nongovernmental organizations, learning institutions, and various government agencies, such as the United States Forest Service.

For the modern era, the U.S. Forest Service has noted three important aspects of the conservation movement: the climate change, water issues, and the education of the public on conservation of the natural environment, especially among children. In regards to climate change, the U.S. Forest Service has undertaken a twenty year research project to develop ways to counteract issues surrounding climate change. However, some small steps have been taken regarding climate change. As rising greenhouse gases contribute to global warming, reforestation projects are seeking to counteract rising carbon emissions.

In Oregon, the Department of Forestry has developed such a small reforestation program in which landowners can lease their land for one hundred years to grow trees. In turn, these trees offset carbon emissions from power companies. Moreover, reforestation projects have other benefits: reforested areas serve as a natural filter of agricultural fertilizers even as new wildlife habitats are created. Reforested land can also contribute to the local economy as rural landowners also distribute hunting leases during the years between harvests.

In essence, projects, such as reforestation, create a viable market of eco-friendly services mutually beneficial to landowners, businesses and society, and most importantly, the environment. Nonetheless, such creative plans will be necessary in the near future as the United States struggles to maintain a positive balance between society and the finite natural resources of the nation. Ultimately, through dedicated research, eco-friendly practices of land management, and efforts to educate the public regarding the necessity of conservation, those individuals dedicated to American conservation seek to preserve the nation's natural resources.

Conservation Issues

Historic:

- Extinction of the Passenger pigeon
- Wholesale hunting of American Bison.

Current:

- Logging of old growth forests
- Mountaintop mining
- Draining and development of the Everglades.

Regulation

Enforcement

Game wardens or conservation officers are employed to protect wildlife and natural areas:

- Louisiana Department of Wildlife & Fisheries-Enforcement Division
- Maryland Department of Natural Resources Police
- Michigan Conservation Officers
- New York State Department of Environmental Conservation Police
- Virginia Department of Game and Inland Fisheries.

Conservation Projects

Restoration of the Everglades

The restoration of the Everglades is an ongoing effort to remedy damage inflicted on the environment of southern Florida during the 20th century. It is the most expensive and comprehensive environmental repair attempt in history. The degradation of the Everglades became an issue in the United States in the early 1970s after a proposal to construct a jetport in the Big Cypress Swamp. Studies indicated the airport would have destroyed the ecosystem in South Florida and Everglades National Park. After decades of destructive practices, both state and federal agencies are looking for ways to balance the needs of the natural environment in South Florida with urban and agricultural centres that have recently and rapidly grown in and near the Everglades.

In response to floods caused by hurricanes in 1947, the Central and Southern Florida Flood Control Project (C&SF) was established to construct flood control devices in the Everglades. The C&SF built 1,400 miles (2,300 km) of canals and levees between the 1950s and 1971 throughout South Florida. Their last venture was the C-38 canal, which straightened the Kissimmee River and caused catastrophic damage to animal habitats, adversely affecting water quality in the region. The canal became the first C&SF project to be reverted when the 22-mile (35 km) canal began to be backfilled, or refilled with the material excavated from it, in the 1980s. The restoration of the Kissimmee River is projected to continue until 2011.

When high levels of phosphorus and mercury were discovered in the waterways in 1986, water quality became a focus for water management agencies. Costly and lengthy court battles were waged between various government entities to determine who was responsible for monitoring and enforcing water quality standards. Governor Lawton Chiles proposed a bill that determined which agencies would have that responsibility, and set deadlines for pollutant levels to decrease in water. Initially the bill was criticized by conservation groups for not being strict enough on polluters, but the Everglades Forever Act was passed in 1994. Since then, the South Florida Water Management District (SFWMD) and the U.S. Army Corps of Engineers have surpassed expectations for achieving lower phosphorus levels.

A commission appointed by Governor Chiles published a report in 1995 stating that South Florida was unable to sustain its growth, and the deterioration of the environment was negatively affecting daily life for residents in South Florida. The environmental decline was predicted to harm tourism and commercial interests if no actions were taken to halt current trends. Results of an eight-year study that evaluated the C&SF were submitted to the United States Congress in 1999. The report warned that if no action was taken the region would rapidly deteriorate. A strategy called the Comprehensive Everglades Restoration Plan was enacted to restore portions of the Everglades, Lake Okeechobee, the Caloosahatchee River, and Florida Bay to undo the damage of the past 50 years. It would take 30 years and cost $7.8 billion to complete. Though the plan was passed into law in 2000, it has been compromised by politics and funding problems.

Background

The Everglades are part of a very large watershed that begins in the vicinity of Orlando. The Kissimmee River drains into Lake Okeechobee, a 730-square-mile (1,900 km^2) lake with an average depth of 9 feet (2.7 m). During the wet season when the lake exceeds its capacity, the water leaves the lake in a very wide and shallow river, approximately 100 miles (160 km) long and 60 miles (97 km) wide. This wide and shallow flow is known as sheetflow. The land gradually slopes toward Florida Bay, the historical destination of most of the water leaving the Everglades. Before drainage attempts, the Everglades comprised 4,000 square miles (10,000 km^2), taking up a third of the Florida peninsula.

Since the early 19th century the Everglades have been a subject of interest for agricultural development. The first attempt to drain the Everglades occurred in 1882 when a Pennsylvania land developer named Hamilton Disston constructed the first canals. Though these attempts were largely unsuccessful, Disston's purchase of land spurred tourism and real estate development of the state. The political motivations of Governor Napoleon Bonaparte Broward resulted in more successful attempts at canal construction between 1906 and 1920. Recently reclaimed wetlands were used for cultivating sugarcane and vegetables, while urban development began in the Everglades. The 1926 Miami Hurricane and the 1928 Okeechobee Hurricane caused widespread devastation and flooding which prompted the Army Corps of Engineers to construct a dike around Lake Okeechobee. The four-story wall cut off water from the Everglades. Floods from hurricanes in 1947 motivated the U.S. Congress to establish the Central and Southern Florida Flood Control Project (C&SF), responsible for constructing 1,400 miles (2,300 km) of canals and levees, hundreds of pumping stations and other water control devices. The C&SF established Water Conservation Areas (WCAs) in 37% of the original Everglades, which acted as reservoirs providing excess water to the South Florida metropolitan area, or flushing it into the Atlantic Ocean or the Gulf of Mexico. The C&SF also established the Everglades Agricultural Area (EAA), which grows the majority of sugarcane crops in the United States. When the EAA was first established, it encompassed approximately 27% of the original Everglades.

By the 1960s, urban development and agricultural use had decreased the size of the Everglades considerably. The remaining 25% of the Everglades in its original state is protected in Everglades National Park, but the park was established before the C&SF, and it depended upon the actions of the C&SF to release water. As Miami and other metropolitan areas began to intrude on the Everglades in the 1960s, political battles took place between park management and the C&SF when insufficient water in the park threw ecosystems into chaos. Fertilizers used in the EAA began to alter soil and hydrology in Everglades National Park, causing the proliferation of exotic plant species. However, a proposition to build a massive jetport in the Big Cypress Swamp in 1969 focused attention on the degraded natural systems in the Everglades. For the first time, the Everglades became a subject of environmental conservation.

Everglades as a Priority

In the 1970s protecting the environment became a national priority. Time magazine declared it the Issue of the Year in January 1971, reporting that it was rated as Americans' "most serious problem confronting their community—well ahead of crime, drugs and poor schools". South Florida became a poignant subject when the region experienced a severe drought from 1970 to 1975: Miami received only 33 inches (840 mm) of rain in 1971—22 inches (560 mm) less than average. With the assistance of governor's aide Nathaniel Reed and U.S. Fish and Wildlife Service biologist Arthur R. Marshall, politicians began to take action on the Everglades. Governor Reubin Askew implemented the Land Conservation Act in 1972, which allowed the state to use voter-approved bonds of $240 million to purchase land considered to be environmentally unique and irreplaceable. Since then, Florida has purchased more land for public use than any other state. In 1972 President Richard Nixon declared the Big Cypress Swamp—the intended location for the Miami jetport in 1969—to be federally protected. Big Cypress National Preserve was established in 1974, and Fakahatchee Strand State Preserve was created the same year.

In 1976, Everglades National Park was declared an International Biosphere Reserve by UNESCO, which also listed the park as a World Heritage Site in 1979. The Ramsar Convention designated the Everglades a Wetland of International Importance in 1987. As such, it became one of only three locations on earth which have appeared on all three lists with Lake Ichkeul in Tunisia and Srebarna Lake in Bulgaria.

Kissimmee River

In the 1960s, the C&SF came under increased scrutiny from government overseers and conservation groups. Critics maintained its size was comparable to the Tennessee Valley Authority's dam-building projects during the Great Depression, and that its projects had run into the billions of dollars without any apparent resolution or plan. The projects of the C&SF have been characterized as part of "crisis and response" cycles that "ignored the consequence for the full system, assumed certainty of the future, and succeeded in solving the momentary crisis, but set in motion conditions that exaggerate future crises". The last project began in 1962: it was to build a canal to straighten the winding floodplain of the Kissimmee River that had historically fed Lake Okeechobee which in turn fed the Everglades. Marjory Stoneman Douglas later wrote that the C&SF projects were "interrelated stupidity", crowned by the C-38 canal. The canal, which aimed to replace a meandering 90-mile (140 km) river with a 52-mile (84 km^2) channel, was completed in 1971 and cost $29 million. It supplanted approximately 45,000 acres (180 km²) of marshland with retention ponds, dams, and vegetation. Loss of habitat has caused the region to experience a drastic loss of waterfowl, wading birds, and game fish. The reclaimed floodplains were taken over by agriculture, bringing fertilizers and insecticides that washed into Lake Okeechobee. Even before the canal was finished, conservation organizations and sport fishing and hunting groups were calling for the restoration of the Kissimmee River.

Arthur R. Marshall led the efforts to undo the damage. According to Douglas, Marshall was successful in portraying the Everglades from the Kissimmee Chain of Lakes to Florida Bay—including the atmosphere, climate, and limestone—as a single organism.

Rather than remaining the preserve of conservation organizations, the cause of restoring the Everglades became a priority for politicians. Douglas observed, "Marshall accomplished the extraordinary magic of taking the Everglades out of the bleeding-hearts category forever". His efforts led to recognition, and Arthur R. Marshall's name would eventually grace a national wildlife refuge in Palm Beach County and an endowed chair at the University of Florida.

At the insistent urging of Marshall, newly elected Governor Bob Graham announced the formation of the "Save Our Everglades" campaign in 1983, and in 1985 Graham lifted the first shovel of backfill for a portion of the C-38 canal. Within a year the area was covered with water returning to its original state. Graham declared that by the year 2000, the Everglades would resemble its predrainage state as much as possible. In 1992, the Kissimmee River Restoration Project was approved by Congress in the Water Resources Development Act. The project was estimated to cost $578 million to convert only 22 miles (35 km) of the canal; the cost was designed to be divided between the state of Florida and the U.S. government, with the state being responsible for purchasing land to be restored. A project manager for the Army Corps of Engineers explained in 2002, "What we're doing on this scale is going to be taken to a larger scale when we do the restoration of the Everglades". The entire project is estimated to be completed by 2011.

Water Quality

Attention to water quality was focused in South Florida in 1986 when a widespread algal bloom occurred in one-fifth of Lake Okeechobee. The bloom was discovered to be the result of fertilizers from the Everglades Agricultural Area. Although laws stated in 1979 that the chemicals used in the EAA should not be deposited into the lake, they were flushed into the canals that fed the Everglades Water Conservation Areas, and eventually pumped into the lake. Microbiologists discovered that, although phosphorus assists plant growth, it destroys periphyton, one of the basic building blocks of marl in the Everglades. Marl is one of two types of Everglades soil, along with peat; it is found where parts of the Everglades are flooded for shorter periods of time as layers of periphyton dry. Most of the phosphorus compounds also rid peat of dissolved oxygen and promote algae growth, causing native invertebrates to die, and sawgrass to be replaced with invasive cattails that grow too tall and thick for birds and alligators to nest in. Tested water showed 500 parts per billion (ppb) of phosphorus near sugarcane fields. State legislation in 1987 mandated a 40% reduction of phosphorus by 1992.

The sugarcane industry, dominated by two companies named U.S. Sugar and Flo-Sun, was responsible for more than half of the crop in the EAA. They were well-represented in state and federal governments by lobbyists who enthusiastically protected their interests. The Audubon Society pointed out that the sugar industry, nicknamed "Big Sugar", donated more money to political parties and candidates than General Motors. The sugar industry attempted to block government-funded studies of polluted water, and when the federal prosecutor in Miami faulted the sugar industry in legal action to protect Everglades National Park, Big Sugar tried to get the lawsuit withdrawn and the prosecutor fired. A costly legal battle ensued from 1988 to 1992 between the State of Florida, the U.S. Government, and the sugar industry to resolve who was responsible for water quality standards, the maintenance of Everglades National Park and the Arthur R. Marshall Loxahatchee National Wildlife Refuge.

A different concern about water quality arose when mercury was discovered in fish during the 1980s. Because mercury is damaging to people, warnings were posted for fishermen that cautioned against eating fish caught in South Florida, and scientists became alarmed when a Florida panther was found dead near Shark River Slough with mercury levels high enough to be fatal to humans. When mercury is ingested it adversely affects the central nervous system, and can cause brain damage and birth defects. Studies of mercury levels found that it is bioaccumulated through the food chain: animals that are lower on the chain have decreased amounts, but as larger animals eat them, the amount of mercury is multiplied. The dead panther's diet consisted of small animals, including raccoons and young alligators. The source of the mercury was found to be waste incinerators and fossil fuel power plants that expelled the element in the atmosphere, which precipitated along with rain, or in the dry season, dust. Naturally occurring bacteria in the Everglades that function to reduce sulfur also transform mercury deposits into methylmercury. This process was more dramatic in areas where flooding was not as prevalent. Because of requirements that reduced power plant and incinerator emissions, the levels of mercury found in larger animals decreased as well: approximately a 60% decrease in fish and a 70% decrease in birds, though some levels still remain a health concern for people.

Everglades Forever Act

In an attempt to resolve the political quagmire over water quality, Governor Lawton Chiles introduced a bill in 1994 to clean up water within the EAA that was being released to the lower Everglades. The bill stated that the "Everglades ecosystem must be restored both in terms of water quality and water quantity and must be preserved and protected in a manner that is long term and comprehensive". It ensured the Florida Department of Environmental Protection (DEP) and the South Florida Water Management District (SFWMD) would be responsible for researching water quality, enforcing water supply improvement, controlling exotic species, and collecting taxes, with the aim of decreasing the levels of phosphorus in the region. It allowed for purchase of land where pollutants would be sent to "treat and improve the quality of waters coming from the EAA".

Critics of the bill argued that the deadline for meeting the standards was unnecessarily delayed until 2006—a period of 12 years—to enforce better water quality. They also maintained that it did not force sugarcane farmers, who were the primary polluters, to pay enough of the costs, and increased the threshold of what was an acceptable amount of phosphorus in water from 10 ppb to 50 ppb. Governor Chiles initially named it the Marjory Stoneman Douglas Act, but Douglas was so unimpressed with the action it took against polluters that she wrote to Chiles and demanded her name be stricken from it. Despite criticism, the Florida legislature passed the Act in 1994. The SFWMD stated that its actions have exceeded expectations earlier than anticipated, by creating Stormwater Treatment Areas (STA) within the EAA that contain a calcium-based substance such as lime rock layered between peat, and filled with calcareous periphyton. Early tests by the Army Corps of Engineers revealed this method reduced phosphorus levels from 80 ppb to 10 ppb. The STAs are intended to treat water until the phosphorus levels are low enough to be released into the Loxahatchee National Wildlife Refuge or other WCAs.

Wildlife Concerns

The intrusion of urban areas into wilderness has had a substantial impact on wildlife, and several species of animals are considered endangered in the Everglades region. One animal that has benefited from endangered species protection is the American Alligator (Alligator mississippiensis), whose holes give refuge to other animals, often allowing many species to survive during times of drought. Once abundant in the Everglades, the alligator was listed as an endangered species in 1967, but a combined effort by federal and state organizations and the banning of alligator hunting allowed it to rebound; it was pronounced fully recovered in 1987 and is no longer an endangered species. However, alligators' territories and average body masses have been found to be generally smaller than in the past, and because populations have been reduced, their role during droughts has become limited.

The American Crocodile (Crocodylus acutus) is also native to the region and has been designated as endangered since 1975. Unlike their relatives the alligators, crocodiles tend to thrive in brackish or salt-water habitats such as estuarine or marine coasts. Their most significant threat is disturbance by people. Too much contact with humans causes females to abandon their nests, and males in particular are often victims of vehicle collisions while roaming over large territories and attempting to cross U.S. 1 and Card Sound Road in the Florida Keys. There are an estimated 500 to 1,000 crocodiles in southern Florida.

The most critically endangered of any animal in the Everglades region is the Florida panther, a species that once lived throughout the southeastern United States: there were only 25–30 in the wild in 1995. The panther is most threatened by urban encroachment, because males require approximately 200 square miles (520 km^2) for breeding territory. A male and two to five females may live within that range. When habitat is lost, panthers will fight over territory. After vehicle collisions, the second most frequent cause of death for panthers is intra-species aggression. In the 1990s urban expansion crowded panthers from southwestern Florida as Naples and Ft. Myers began to expand into the western Everglades and Big Cypress Swamp. Agencies such as the Army Corps of Engineers and the U.S. Fish and Wildlife Service were responsible for maintaining the Clean Water Act and the Endangered Species Act, yet still approved 99% of all permits to build in wetlands and panther territory. A limited genetic pool is also a danger. Biologists introduced eight female Texas cougars to diversify genes, and there are between 80 and 100 panthers in the wild as of 2008.

Perhaps the most dramatic loss of any group of animals has been to wading birds. Their numbers were estimated by eyewitness accounts to be approximately 2.5 million in the late 19th century. However, snowy egrets (Egretta thula), roseate spoonbills (Platalea ajaja), and reddish egrets (Egretta rufescens) were hunted to the brink of extinction for the colourful feathers used in women's hats. After about 1920 when the fashion passed, their numbers returned in the 1930s, but over the next 50 years actions by the C&SF further disturbed populations. When the canals were constructed, natural water flow was restricted from the mangrove forests near the coast of Florida Bay. From one wet season to the next, fish were unable to reach traditional locations to repopulate when water was withheld by the C&SF. Birds were forced to fly farther from their nests to forage for food. By the 1970s, bird numbers had decreased 90%. Many of the birds moved to smaller colonies in the WCAs to be closer to a food source, making them more difficult to count. Yet they remain significantly fewer in number than before the canals were constructed.

Invasive Species

Around 6 million people moved to South Florida between 1940 and 1965. With a thousand people moving to Miami each week, urban development quadrupled. As the human population grew rapidly, the problem of exotic plant and animal species also grew. Many species of plants were brought in to South Florida from Asia, Central America, or Australia as decorative landscaping. Exotic animals imported by the pet trade have escaped or been released. Biological controls that keep invasive species smaller in size and fewer in number in their native lands often do not exist in the Everglades, and they compete with the embattled native species for food and space. Of imported plant species, melaleuca trees (Melaleuca quinquenervia) have caused the most problems. Melaleucas grow on average 100 feet (30 m) in the Everglades, as opposed to 25 to 60 feet (7.6 to 18 m) in their native Australia. They were brought to southern Florida as windbreaks and deliberately seeded in marsh areas because they absorb vast amounts of water. In a region that is regularly shaped by fire, melaleucas are fire-resistant and their seeds are more efficiently spread by fire. They are too dense for wading birds with large wingspans to nest in, and they choke out native vegetation. Costs of controlling melaleucas topped $2 million in 1998 for Everglades National Park. In Big Cypress National Preserve, melaleucas covered 186 square miles (480 km^2) at their most pervasive in the 1990s.

Brazilian pepper (Schinus terebinthifolius) was brought to Southern Florida as an ornamental shrub and was dispersed by the droppings of birds and other animals that ate its bright red berries. It thrives on abandoned agricultural land growing in forests too dense for wading birds to nest in, similar to melaleucas. It grows rapidly especially after hurricanes and has invaded pineland forests. Following Hurricane Andrew, scientists and volunteers cleared damaged pinelands of Brazilian pepper so the native trees would be able to return to their natural state.

The species that is causing the most impediment to restoration is the Old World climbing fern (Lygodium microphyllum), introduced in 1965. The fern grows rapidly and thickly on the ground, making passage for land animals such as black bears and panthers problematic. The ferns also grow as vines into taller portions of trees, and fires climb the ferns in "fire ladders" to scorch portions of the trees that are not naturally resistant to fire.

Several animal species have been introduced to Everglades waterways. Many tropical fish are released, the most detrimental being the blue tilapia (Oreochromis aureus), which builds large nests in shallow waters. Tilapia also consume vegetation which would normally be used by young native fishes for cover and protection.

Reptiles have a particular affinity for the South Florida ecosystem. Virtually all lizards appearing in the Everglades have been introduced, such as the brown anole (Anolis sagrei) and the tropical house gecko (Hemidactylus mabouia). The herbivorous green iguana (Iguana iguana) can reproduce rapidly in wilderness habitats. However, the reptile that has earned media attention for its size and potential to harm children and domestic pets is the Burmese python (Python molurus bivittatus), which has spread quickly throughout the area. The python can grow up to 20 feet (6.1 m) long and competes with alligators for the top of the food chain.

Though exotic birds such as parrots and parakeets are also found in the Everglades, their impact is negligible. Conversely, perhaps the animal that causes the most damage to native wildlife is the domestic or feral cat. Across the U.S., cats are responsible for approximately a billion bird deaths annually. They are estimated to number 640 per square mile; cats living in suburban areas have devastating effects on migratory birds and marsh rabbits.

Homestead Air Force Base

Hurricane Andrew struck Miami in 1992, with catastrophic damage to Homestead Air Force Base in Homestead. A plan to rejuvenate the property in 1993 and convert it into a commercial airport was met with enthusiasm from local municipal and commercial entities hoping to recoup $480 million and 11,000 jobs lost in the local community by the destruction and subsequent closing of the base. On March 31, 1994, the base was designated as an Air Reserve Base, functioning only part-time. A cursory environmental study performed by the Air Force was deemed insufficient by local conservation groups, who threatened to sue in order to halt the acquisition when estimates of 650 flights a day were projected. Groups had previously been alarmed in 1990 by the inclusion of Homestead Air Force Base on a list of the U.S. Government's most polluted properties. Their concerns also included noise, and the inevitable collisions with birds using the mangrove forests as rookeries. The air force base is located between Everglades National Park and Biscayne National Park, giving it the potential to cause harm to both. In 2000, Secretary of the Interior Bruce Babbitt and the director of the U.S. Environmental Protection Agency expressed their opposition to the project, despite other Clinton Administration agencies previously working to ensure the base would be turned over to local agencies quickly and smoothly as "a model of base disposal". Although attempts were made to make the base more environmentally friendly, in 2001 local commercial interests promoting the airport lost federal support.

Comprehensive Everglades Restoration Plan

Sustainable South Florida

Despite the successes of the Everglades Forever Act and the decreases in mercury levels, the focus intensified on the Everglades in the 1990s as quality of life in the South Florida metropolitan areas diminished. It was becoming clear that urban populations were consuming increasingly unsustainable levels of natural resources. A report entitled "The Governor's Commission for a Sustainable South Florida", submitted to Lawton Chiles in 1995, identified the problems the state and municipal governments were facing. The report remarked that the degradation of the natural quality of the Everglades, Florida Bay, and other bodies of water in South Florida would cause a significant decrease in tourism (12,000 jobs and $200 million annually) and income from compromised commercial fishing

(3,300 jobs and $52 million annually). The report noted that past abuses and neglect of the environment had brought the region to "a precipitous juncture" where the inhabitants of South Florida faced health hazards in polluted air and water; furthermore, crowded and unsafe urban conditions hurt the reputation of the state. It noted that though the population had increased by 90% over the previous two decades, registered vehicles had increased by 166%. On the quality and availability of water, the report stated, "[The] frequent water shortages ... create the irony of a natural system dying of thirst in a subtropical environment with over 53 inches of rain per year".

Restoration of the Everglades, however, briefly became a bipartisan cause in national politics. A controversial penny-a-pound (2 cent/kg) tax on sugar was proposed to fund some of the necessary changes to be made to help decrease phosphorus and make other improvements to water. State voters were asked to support the tax, and environmentalists paid $15 million to encourage the issue. Sugar lobbyists responded with $24 million in advertising to discourage it and succeeded; it became the most expensive ballot issue in state history. How restoration might be funded became a political battleground and seemed to stall without resolution. However, in the 1996 election year, Republican senator Bob Dole proposed that Congress give the State of Florida $200 million to acquire land for the Everglades. Democratic Vice President Al Gore promised the federal government would purchase 100,000 acres (400 km^2) of land in the EAA to turn it over for restoration. Politicking reduced the number to 50,000 acres (200 km^2), but both Dole's and Gore's gestures were approved by Congress.

Central and South Florida Project Restudy

As part of the Water Resources Development Act of 1992, Congress authorized an evaluation of the effectiveness of the Central and Southern Florida Flood Control Project. A report known as the "Restudy", written by the U.S. Army Corps of Engineers and the South Florida Water Management District, was submitted to Congress in 1999. It cited indicators of harm to the system: a 50% reduction in the original Everglades, diminished water storage, harmful timing of water release, an 85 to 90% decrease in wading bird populations over the past 50 years, and the decline of output from commercial fisheries.

Bodies of water including Lake Okeechobee, the Caloosahatchee River, St. Lucie estuary, Lake Worth Lagoon, Biscayne Bay, Florida Bay, and the Everglades reflected drastic water level changes, hypersalinity, and dramatic changes in marine and freshwater ecosystems. The Restudy noted the overall decline in water quality over the past 50 years was caused by loss of wetlands that act as filters for polluted water. It predicted that without intervention the entire South Florida ecosystem would deteriorate. Canals took roughly 170 billion US gallons (640 Gl) of water to the Atlantic Ocean or Gulf of Mexico daily, so there was no opportunity for water storage, yet flooding was still a problem. Without changes to the current system, the Restudy predicted water restrictions would be necessary every other year, and annually in some locations. It also warned that revising some portions of the project without dedicating efforts to an overall comprehensive plan would be insufficient and probably detrimental.

After evaluating ten plans, the Restudy recommended a comprehensive strategy that would cost $7.8 billion over 20 years. The plan advised taking the following actions:

- Create surface water storage reservoirs to capture 1,500,000 acre feet (1.85 km^3) of water in several locations taking up 181,300 acres (734 km^2).
- Create water preserve areas between Miami-Dade and Palm Beach and the eastern Everglades to treat runoff water.
- Manage Lake Okeechobee as an ecological resource to avoid the drastic rise and fall of water levels in the lake that are harmful to aquatic plant and animal life and disturb the lake sediments.
- Improve water deliveries to estuaries to reduce the rapid discharge of excess water to the Caloosahatchee and St. Lucie estuaries that upset nutrient balances and cause lesions on fish. Stormwater discharge would be sent instead to reservoirs.
- Increase underground water storage to hold 16 billion US gallons (61 Gl) a day in wells, or reservoirs in the Floridan Aquifer, to be used later in dry periods, in a method called Aquifer Storage and Recovery (ASR).

- Construct treatment wetlands as Stormwater Treatment Areas throughout 35,600 acres (144 km^2), that would decrease the amount of pollutants in the environment.
- Improve water deliveries to the Everglades by increasing them at a rate of approximately 26% into Shark River Slough.
- Remove barriers to sheetflow by destroying or removing 240 miles (390 km) of canals and levees, specifically removing the Miami Canal and reconstructing the Tamiami Trail from a highway to culverts and bridges to allow sheetflow to return to a more natural rate of water flow into Everglades National Park.
- Store water in quarries and reuse wastewater by employing existing quarries to supply the South Florida metropolitan area as well as Florida Bay and the Everglades. Construct two wastewater treatment plants capable of discharging 22 billion US gallons (83 Gl) a day to recharge the Biscayne Aquifer.

The implementation of all of the advised actions, the report stated, would "result in the recovery of healthy, sustainable ecosystems throughout south Florida". The report admitted that it did not have all the answers, though no plan could. However, it predicted that it would restore the "essential defining features of the pre-drainage wetlands over large portions of the remaining system", that populations of all animals would increase, and animal distribution patterns would return to their natural states. Critics expressed concern over some unused technology; scientists were unsure if the quarries would hold as much water as was being suggested, and whether the water would harbour harmful bacteria from the quarries. Overtaxing the aquifers was another concern—it was not a technique that had been previously attempted.

It is important to understand that the 'restored' Everglades of the future will be different from any version of the Everglades that has existed in the past. While it certainly will be vastly superior to the current ecosystem, it will not completely match the pre-drainage system. This is not possible, in light of the irreversible physical changes that have made (sic) to the ecosystem. It will be an Everglades that is smaller and somewhat differently arranged than the historic ecosystem. But it will be a successfully restored Everglades, because it will have recovered those hydrological and biological patterns which defined the original Everglades, and which made it unique among the world's wetland systems. It will become a place that kindles the wildness and richness of the former Everglades.

The report was the result of many cooperating agencies that often had conflicting goals. An initial draft was submitted to Everglades National Park management who asserted not enough water would be released to the park quickly enough—that the priority went to delivering water to urban areas. When they threatened to refuse to support it, the plan was rewritten to provide more water to the park.

However, the Miccosukee Indians have a reservation in between the park and water control devices, and they threatened to sue to ensure their tribal lands and a $50 million casino would not be flooded. Other special interests were also concerned that businesses and residents would take second priority after nature. The Everglades, however, proved to be a bipartisan cause. The Comprehensive Everglades Restoration Plan (CERP) was authorized by the Water Development Act of 2000 and signed into law by President Bill Clinton on December 11, 2000. It approved the immediate use of $1.3 billion for implementation to be split by the federal government and other sources.

Implementation

The State of Florida reports that it has spent more than $2 billion on the various projects since CERP was signed. More than 36,000 acres (150 km^2) of Stormwater Treatment Areas (STA) have been constructed to filter 2,500 short tons (2,300 t) of phosphorus from Everglades waters. An STA covering 17,000 acres (69 km^2) was constructed in 2004, making it the largest manmade wetland in the world. Fifty-five percent of the land necessary for restoration, totalling 210,167 acres (850.5 km^2), has been purchased by the State of Florida. A plan named "Acceler 8", to hasten the construction and funding of the project, was put into place, spurring the start of six of eight construction projects, including that of three large reservoirs.

Despite the bipartisan goodwill and declarations of the importance of the Everglades, the region still remains in danger. Political maneuvering continues to impede CERP: sugar lobbyists promoted a bill in the Florida legislature in 2003 that increased the acceptable amount of phosphorus in Everglades waterways from 10 ppb to 15 ppb and extended the deadline for the mandated decrease by 20 years. A compromise of 2016 was eventually reached. Environmental organizations express concern that attempts to speed up some of the construction through Acceler 8 are politically motivated; the six projects Acceler 8 focuses on do not provide more water to natural areas in desperate need of it, but rather to projects in populated areas bordering the Everglades, suggesting that water is being diverted to make room for more people in an already overtaxed environment. Though Congress promised half the funds for restoration, after the War in Iraq began and two of CERP's major supporters in Congress retired, the federal role in CERP was left unfulfilled. According to a story in The New York Times, state officials say the restoration is lost in a maze of "federal bureaucracy, a victim of 'analysis paralysis' ". In 2007, the release of $2 billion for Everglades restoration was approved by Congress, overriding President George W. Bush's veto of the entire Water Development Project the money was a part of. Bush's rare veto went against the wishes of Florida Republicans, including his brother, Governor Jeb Bush. A lack of subsequent action by the Congress prompted Governor Charlie Crist to travel to Washington D.C. in February 2008 and inquire about the promised funds. As of June 2008, the federal government has spent only $400 million of the $7.8 billion legislated. Carl Hiassen characterized George W. Bush's attitude toward the environment as "long-standing indifference" in June 2008, exemplified when Bush stated he would not intervene to change the Environmental Protection Agency's (EPA) policy allowing the release of water polluted with fertilizers and phosphorus into the Everglades.

Reassessment of CERP

Florida still receives a thousand new residents daily and lands slated for restoration and wetland recovery are often bought and sold before the state has a chance to bid on them. The competitive pricing of real estate also drives it beyond the purchasing ability of the state. Because the State of Florida is assisting with purchasing lands and funding construction, some of the programs under CERP are vulnerable to state budget cuts. In June 2008 Governor Crist announced that the State of Florida will buy U.S. Sugar for $1.7 billion. The idea came when sugar lobbyists were trying to persuade Crist to relax restriction of U.S. Sugar's practice of pumping phosphorus-laden water into the Everglades. According to one of the lobbyists who characterized it as a "duh moment", Crist said, "If sugar is polluting the Everglades, and we're paying to clean the Everglades, why don't we just get rid of sugar?" The largest producer of cane sugar in the U.S. will continue operations for six years, and when ownership transfers to Florida, 187,000 acres (760 km²) of the Everglades will remain undeveloped to allow it to be restored to its pre-drainage state.

In September 2008 the National Research Council (NRC), a nonprofit agency providing science and policy advice to the federal government, submitted a report on the progress of CERP. The report noted "scant progress" in restoration because of problems in budgeting, planning, and bureaucracy. The NRC report called the Everglades one of the "world's treasured ecosystems" that is being further endangered by lack of progress: "Ongoing delay in Everglades restoration has not only postponed improvements—it has allowed ecological decline to continue". It cited the shrinking tree islands, and the negative population growth of the endangered Rostrhamus sociabilis or Everglades snail kite, and Ammodramus maritimus mirabilis, the Cape Sable seaside sparrow. The lack of water reaching Everglades National Park was characterized as "one of the most discouraging stories" in implementation of the plan. The NRC recommended improving planning on the state and federal levels, evaluating each CERP project annually, and further acquisition of land for restoration. Everglades restoration was earmarked $96 million in federal funds as part of the American Recovery and Reinvestment Act of 2009 with the intention of providing civil service and construction jobs while simultaneously implementing the legislated repair projects.

In January 2010, work began on the C-111 canal, built in the 1960s to drain irrigated farmland, to reconstruct it to keep from diverting water from Everglades National Park. Two other projects focusing on restoration are also scheduled to start in 2010. Governor Crist announced the same month that $50 million would be earmarked for Everglades restoration. In April of the same year, a federal district court judge sharply criticized both state and federal failures to meet deadlines, describing the cleanup efforts as being slowed by "glacial delay" and government neglect of environmental law enforcement "incomprehensible".

Protected Areas

The protected areas of the United States are managed by an array of different federal, state, tribal and local level authorities and receive widely varying levels of protection. Some areas are managed as wilderness, while others are operated with acceptable commercial exploitation. As of 31 January 2008 (2008-01-31), according to the United Nations Environment Programme, the U.S. had a total of 6,770 terrestrial nationally designated (federal) protected areas. These protected areas cover 2,607,131 km² (1,006,619 sq mi), or 27.08 percent of the land area of the United States. This is also one-tenth of the protected land area of the world. The U.S. also had a total of 787 National Marine Protected Areas, covering an additional 627,830 km² (242,410 sq mi), or 67 percent of the total marine area of the United States. In addition, the World Commission on Protected Areas' 2009 database has over 10,480 protected areas listed for the U.S., including the state level protected areas.

Some areas are managed in concert between levels of government. The Father Marquette National Memorial is an example of a federal park operated by a state park system, while Kal-Haven Trail is an example of a state park operated by county-level government.

Federal Level Protected Areas

Federal level protected areas are managed by a variety of agencies, most of which are a part of the National Park Service, a bureau of the United States Department of the Interior. They are often considered the crown jewels of the protected areas. Other areas are managed by the United States Forest Service, the Bureau of Land Management and the United States Fish and Wildlife Service. The United States Army Corps of

Engineers is claimed to provide 30 percent of the recreational opportunities on federal lands, mainly through lakes and waterways that they manage.

The highest levels of protection, as described by the International Union for Conservation of Nature (IUCN), are Level I (Strict Nature Reserves & Wilderness Areas) and Level II (National Parks). The United States maintains 12 percent of the Level I and II lands in the world. These lands had a total area of 210,000 sq mi (540,000 km²).

A confusing system for naming protected areas results in some types being used by more than one agency. For instance, both the National Park Service and the U.S. Forest Service operate areas designated National Preserves and National Recreation Areas. The National Park Service, the U.S. Forest Service, and the Bureau of Land Management operate areas called National Monuments. National Wilderness Areas are designated within other protected areas, managed by various agencies and sometimes wilderness areas span areas managed by multiple agencies.

There are existing Federal designations of historic or landmark status that may support preservation via tax incentives, but that do not necessarily convey any protection, including a listing on the National Register of Historic Places or a designation as a National Historic Landmark. States and local zoning bodies may or may not choose to protect these. The state of Colorado, for example, is very clear that it does not set any limits on owners of NRHP properties:

- List of areas in the United States National Park System
- List of National Parks of the United States
- List of biosphere reserves in the United States
- List of National Wild and Scenic Rivers
- List of National Wildlife Refuges
- List of U.S. National Forests, including United States National Grasslands
    - List of largest National Forests
- List of U.S. National Marine Sanctuaries
- List of U.S. Wilderness Areas
- National Landscape Conservation System.

State Level Protected Areas

Every state has a system of state parks. State parks vary widely from urban parks to very large parks that are on a par with national parks. Some state parks, like Adirondack Park, are similar to the National parks of England and Wales, with numerous towns inside the borders of the park. About half the area of the park, some 3,000,000 acres (1,200,000 ha), is state-owned and preserved as "forever wild" by the Forest Preserve of New York. Wood-Tikchik State Park in Alaska claims to be the largest state park by the amount of contiguous protected land; it is larger than many U.S. National Parks, with some 1,600,000 acres (650,000 ha). Many states also operate game and recreation areas.

Local Level Protected Areas

Various counties, cities, metropolitan authorities, regional parks, townships, soil conservation districts and other units manage a variety of local level parks. Some of these are little more than picnic areas or playgrounds; however, others are extensive nature reserves. South Mountain Park in Phoenix, Arizona, for example, is called the largest city park in the United States; it spans 25 sq mi (65 km^2) and contains 58 mi (93 km) of trails.

3

United States—Transportation and Trade

Historically, societies have always located near water, due partly to the fact that water enables more efficient travel compared to going over land. Waterways are critically important to the transportation of people and goods throughout the world. The complex network of connections between coastal ports, inland ports, rail, air, and truck routes forms a foundation of material economic wealth worldwide.

Within the United States, waterways have been developed and integrated into a world-class transportation system that has been instrumental in the country's economic development. Today, there are more than 17,700 kilometers of commercially important navigation channels in the lower 48 states.

Early History of Water-based Transportation

The historical development of water-based transportation is connected to the importance of domestic and international trade. Early exploration of North America identified large amounts of natural resources such as fisheries, timber, and furs. Trade centres were established along the east coast of North America where goods could be gathered together and ocean vessels could transport them to consumers in Europe and other foreign areas. The success of commercial trading companies spurred the introduction of

From the sixteenth to the eighteenth centuries, small subsistence farms were prevalent among the American colonies.

Eventually larger farms emerged and produced crops such as wheat, tobacco, rice, indigo, and cotton that were commercially marketable in Europe. Ocean vessels transported the bulk, low-value goods from the colonies to Europe and returned with high-value, low-density goods such as inks, linens, and finished products that had a much higher return on the investment per vessel trip.

Agricultural production continued to grow and support the growing colonies' economic development. The speed and low cost of transporting goods by water influenced the locations of population settlements near navigable water (rivers, lakes, canals, and oceans). Goods produced on inland farms were transported via inland waterways to the coastal ports. Goods shipped by smaller vessels from surrounding ports were transported to New York, Boston, and Philadelphia, and exported on larger oceangoing ships. These ships from the smaller ports then transported imported goods back to the surrounding ports.

During the 1700s, the British government passed many acts, such as the Navigation Acts and the Stamp Act of 1765, designed to collect taxes from the colonists. The acts affected trade, and were met with opposition from the colonist. In Philadelphia during the fall of 1774, the "Declarations and Resolves of the First Continental Congress" called for non-importation of British goods, and became a catalyst for the American Revolutionary War (1775–1784). The resulting independence for the United States allowed trade a free rein, and it flourished.

Westward Expansion

The westward expansion of the United States exposed a wealth of natural resources and an increased production in agricultural goods. The inland transportation infrastructure of roads, railroads, canals, and rivers connected the early western settlers with the rest of the nation, and enabled goods to move from the west back to more populated areas in the east and onto other parts of the world. The River and Harbour (Appropriations) Act of 1876 established federal funding of waterways to promote national commerce but not to benefit any particular state nor to allow waterway tolls.

Twentieth and Twenty-First Centuries

Increased levels of world trade resulted from the economic growth occurring since the end of World War II in 1945. The United States was in the position to take advantage of new trading opportunities as new world markets opened. Developing countries demanded capital goods, agricultural products, consumer goods, and commercial services, which the United States could provide. As these nations produced goods for export, the United States became a market for these goods.

A significant factor in the opening of the inland waterway system (and the resultant world trade superiority of the United States) was the advances in ship technology and the application of steam power to ships that travelled the extensive water network. Larger and faster ships emerged from the advances in ship and engine design and improvements in construction materials.

Methods of cargo handling evolved to keep pace with the larger vessel sizes. The introduction of palletization and roll-on/roll-off cargoes enabled vessels to be loaded and unloaded in less time. The emergence of containerization in the late 1950s dramatically affected the shipping industry and port infrastructure. The increasing size of containerized cargo vessels became a driving force in the demand for expanded ports and improved facilities.

Importance to Foreign Trade

Transporting goods with foreign trading partners can be accomplished by road, water, rail, or air. However, for the

About 95 percent of U.S. foreign trade passes through its port system. Ports function as the transfer point between land and water transportation of cargo. For vessels to transport the foreign traded cargo, they must be able to access the ports through established channels. The channels provide adequate water depths for the vessels and navigational aids.

Today's Global Trade

Today the world economy has become globalized. The economic system is changing from one with distinct local and national markets, separated by trade barriers, distance, time, and culture, to one that is increasingly converging and integrating into a global economy. According to the National Oceanic and Atmospheric Administration (NOAA), the United States was the world's leading trader in 1998, accounting for about one billion tons of ocean-bound trade (about 20 percent of the world's total ocean-bound trade) out of about 2.4 billion tons of total foreign trade. In 2000, according to the U.S. Department of Transportation, approximately $736 billion of goods (about 40 percent of the total U.S. foreign trade by dollar amount) were shipped via ocean vessels and passed through U.S. ports. By 2020, international trade is estimated to more than double (by weight) within the United States, with the majority of this trade projected to move via ocean shipping.

Marine Transportation System

According to the Department of Transportation, when cargo is transported by water within the United States, 95 percent of the time it involves the Marine Transportation System (MTS). This comprehensive system resulted from years of water transport development involving such U.S. organizations as the Coast Guard, Maritime Administration, Army Corps of Engineers, NOAA, and Environmental Protection Agency.

The MTS is a complex and diverse national network of waterways, ports, and intermodal landside connections that allows various modes of water-based transportation. The system includes: navigable waterways (such as the Great Lakes-St. Lawrence Seaway); publicly and privately owned commerce-carrying vessels; over 3,500 bulk oil transfer facilities; more than 350 ports located at approximately 4,000 marine terminals; about 40,000 kilometers of navigable channels; more than 235 locks and dams at over 190 locations; shipyards; rail yards; vessel repair facilities, over 10,000 recreational marinas; and a trained labour staff that operates and maintains the entire infrastructure. Users of the waterway system each year include 70,000 port calls for commercial vessels, 110,000 fishing vessels, and 20 million recreational vessels.

International Maritime Fleets and Law

Many nations around the world have built up their fleets to become very profitable. Since World War II, the size of the U.S. flag merchant fleet has declined, partly due to improved technologies and partly due to foreign competition among fleets. But while this number of ships has declined, the productivity has greatly increased. Since 1970, these fewer ships carry 42 percent more cargo. However, the U.S. fleet accounts for less than 5 percent of all commercial foreign trade by weight. Data from the U.S. Army Corps of Engineers and the Maritime Administration indicate the following composition and carrying capacity of the U.S.-flag fleet in 2000:

Composition Capacity

Passenger: 1,265 Passenger: 368,000 passengers

Dry Bulk: 10 Dry Bulk: 2,124,000 metric tons

Dry Cargo/Offshore Support: Dry Cargo: 47,253,000 metric tons 2,910

Containership: 61——

Tanker: 173 Tanker: 19,172,000 metric tons

Vehicular/Railroad Car Ferry: Railroad/Car Float: 89,000 metric 229 tons

Towboat: 5,098——

Dredge: 570 Other [2]: 1,072,000 metric tons

Other [2]: 45 Total: 72,078,000 metric tons/ 368,000 passengers

[1] Includes ferries and day excursion vessels

[2] Includes certain general cargo, roll-on/roll-off, multipurpose, LASH (Lighter Aboard Ship) vessels, and deck barges.

All ships must be registered to one of the world's nations so that responsibility for violations of international laws and conventions may be assigned. This causes many shipping companies to shop around for nations that give them the best values on taxes, wages, and legal restrictions.

Liberia has the largest shipping fleet in the world. Relatively smaller countries like the Bahamas, Honduras, the Marshall Islands, Panama, and Vanuatu have large fleets as well. The United Nation's International Maritime Organization (IMO) is responsible for improving the safety of international shipping, preventing marine pollution, and facilitating international maritime traffic. The Department of Transportation has the overall lead on all maritime issues for the United States, and works with the IMO on these issues.

Economics, National Security, and the Environment

The United States dependence on seas and waterways has been vital to its economic success and national security. The pool of skilled labour working on U.S. flag vessels is also a national security asset. This workforce is relied upon to meet surges in shipping needs in the advent of emergencies. The merchant marine has played a historical role in military conflicts. In 1996, the Maritime Security Act established the maritime security program to support a fleet of U.S. commercial vessels with American crews to support the military and economic security of the country; approximately 47 vessels participate in this program.

The inland waterways are also a national security asset. The 1920 Jones Act required domestic waterborne commerce to be transported in vessels built in the United States, documented under U.S. laws, and owned by U.S. citizens. The Jones Act covers over 42,000 commercial vessels, 124,000 jobs, and $15 billion in economic activity. Many other countries have similar laws restricting foreign access to domestic trade shipped via waterways.

The MTS is also vital to national security. The ability to rapidly deploy troops and materials worldwide is critical to the country's defence. The Voluntary Intermodal Sealift Agreement (VISA) is a standby agreement intended to make commercial, intermodal dry cargo capacity and supporting infrastructure available to meet contingency deployment needs of the Department of Defence. Since World War II, approximately 95 percent of all military equipment and material sent to combat and crisis areas was ship cargo transported by ocean vessels. For example, during the Persian Gulf War (1990–1991), nearly all domestic supplies intended for U.S. military forces travelled by ship.

Marine transportation is an important use of the ocean. Increased demands will be placed on U.S. ports and waterways as domestic and international trade continues to expand. These increases in the use of waterways and port facilities must be achieved while still protecting human health and the environment.

World Transportation—By Land, Sea and Air

Transport or transportation is the movement of people and goods from one location to another. Modes of transport include air, rail, road, water, cable, pipeline, and space. The field can be divided into infrastructure, vehicles, and operations.

Transport infrastructure consists of the fixed installations necessary for transport, and may be roads, railways, airways, waterways, canals and pipelines, and terminals such as airports, railway stations, bus stations, warehouses, trucking terminals, refueling depots (including fueling docks and fuel stations), and seaports. Terminals may be used both for interchange of passengers and cargo and for maintenance.

Vehicles travelling on these networks may include automobiles, bicycles, buses, trains, trucks, people, helicopters, and aircraft. Operations deal with the way the vehicles are operated, and the procedures set for this purpose including financing, legalities and policies. In the transport industry, operations and ownership of infrastructure can be either public or private, depending on the country and mode.

Passenger transport may be public, where operators provide scheduled services, or private. Freight transport has become focused on containerization, although bulk transport is used for large volumes of durable items. Transport plays an important part in economic growth and globalization, but most types cause air pollution and use large amounts of land. While it is heavily subsidized by governments, good planning of transport is essential to make traffic flow, and restrain urban sprawl.

Mode

A mode of transport is a solution that makes use of a particular type of vehicle, infrastructure and operation. The transport of a person or of cargo may involve one mode or several modes, with the latter case being called intermodal or multimodal transport. Each mode has its advantages and disadvantages, and will be chosen for a trip on the basis of cost, capability, route, and speed.

Human-powered

Human-powered transport is the transport of people and/or goods using human muscle-power, in the form of walking, running and swimming. Modern technology has allowed machines to enhance human-power. Human-powered transport remains popular for reasons of cost-saving, leisure, physical exercise and environmentalism. Human-powered transport is sometimes the only type available, especially in underdeveloped or inaccessible regions. It is considered an ideal form of sustainable transportation.

Although humans are able to walk without infrastructure, the transport can be enhanced through the use of roads, especially when enforcing the human power with vehicles, such as bicycles and inline skates. Human-powered vehicles have also been developed for difficult environments, such as snow and water, by watercraft rowing and skiing; even the air can be entered with human-powered aircraft.

Animal-powered

Animal-powered transport is the use of working animals for the movement of people and goods. Humans may ride some of the animals directly, use them as pack animals for carrying goods, or harness them, alone or in teams, to pull sleds or wheeled vehicles. Animals are superior to people in their speed, endurance and carrying capacity; prior to the Industrial Revolution they were used for all land transport impracticable for people, and they remain an important mode of transport in less developed areas of the world.

Air

A fixed-wing aircraft, commonly called airplane, is a heavier-than-air craft where movement of the air in relation to the wings is used to generate lift. The term is used to distinguish from rotary-wing aircraft, where the movement of the lift surfaces relative to the air generates lift. A gyroplane is both fixed-wing and rotary-wing. Fixed-wing aircraft range from small trainers and recreational aircraft to large airliners and military cargo aircraft.

Two things necessary for aircraft are air flow over the wings for lift and an area for landing. The majority of aircraft also need an airport with the infrastructure to receive maintenance, restocking, refuelling and for the loading and unloading of crew, cargo and passengers. While the vast majority of aircraft land and take off on land, some are capable of take off and landing on ice, snow and calm water.

The aircraft is the second fastest method of transport, after the rocket. Commercial jets can reach up to 875 kilometres per hour (544 mph), single-engine aircraft 175 kilometres per hour (109 mph). Aviation is able to quickly transport people and limited amounts of cargo over longer distances, but incur high costs and energy use; for short distances or in inaccessible places helicopters can be used. WHO estimates that up to 500,000 people are on planes at any time.

Rail

Rail transport is where a train runs along a set of two parallel steel rails, known as a railway or railroad. The rails are anchored perpendicular to ties (or sleepers) of timber, concrete or steel, to maintain a consistent distance apart, or gauge. The rails and perpendicular beams are placed on a foundation made of concrete, or compressed earth and gravel in a bed of ballast. Alternative methods include monorail and maglev.

A train consists of one or more connected vehicles that run on the rails. Propulsion is commonly provided by a locomotive, that hauls a series of unpowered cars, that can carry passengers or freight. The locomotive can be powered by steam, diesel or by electricity supplied by trackside systems. Alternatively, some or all the cars can be powered, known as a multiple unit. Also, a train can be powered by horses, cables, gravity, pneumatics and gas turbines. Railed vehicles move with much less friction than rubber tires on paved roads, making trains more energy efficient, though not as efficient as ships.

Intercity trains are long-haul services connecting cities; modern high-speed rail is capable of speeds up to 350 km/h (220 mph), but this requires specially built track. Regional and commuter trains feed cities from suburbs and surrounding areas, while intra-urban transport is performed by high-capacity tramways and rapid transits, often making up the backbone of a city's public transport. Freight trains traditionally used box cars, requiring manual loading and unloading of the cargo. Since the 1960s, container trains have become the dominant solution for general freight, while large quantities of bulk are transported by dedicated trains.

Road

A road is an identifiable route, way or path between two or more places. Roads are typically smoothed, paved, or otherwise prepared to allow easy travel; though they need not be, and historically many roads were simply recognizable routes without any formal construction or maintenance. In urban areas, roads may pass through a city or village and be named as streets, serving a dual function as urban space easement and route.

The most common road vehicle is the automobile; a wheeled passenger vehicle that carries its own motor. Other users of roads include buses, trucks, motorcycles, bicycles and pedestrians. As of 2002, there were 590 million automobiles worldwide.

Automobiles offer high flexibility and with low capacity, but are deemed with high energy and area use, and the main source of noise and air pollution in cities; buses allow for more efficient travel at the cost of reduced flexibility. Road transport by truck is often the initial and final stage of freight transport.

Water

Water transport is the process of transport a watercraft, such as a barge, boat, ship or sailboat, makes over a body of water, such as a sea, ocean, lake, canal or river. The need for buoyancy unites watercraft, and makes the hull a dominant aspect of its construction, maintenance and appearance.

In the 1800s the first steam ships were developed, using a steam engine to drive a paddle wheel or propeller to move the ship. The steam was produced using wood or coal. Now most ships have an engine using a slightly refined type of petroleum called bunker fuel. Some ships, such as submarines, use nuclear power to produce the steam. Recreational or educational craft still use wind power, while some smaller craft use internal combustion engines to drive one or more propellers, or in the case of jet boats, an inboard water jet. In shallow draft areas, hovercraft are propelled by large pusher-prop fans. Although slow, modern sea transport is a highly effective method of transporting large quantities of non-perishable goods. Commercial vessels, nearly 35,000 in number, carried 7.4 billion tons of cargo in 2007. Transport by water is significantly less costly than air transport for trans-continental shipping; short sea shipping and ferries remain viable in coastal areas.

Other

Pipeline transport sends goods through a pipe, most commonly liquid and gases are sent, but pneumatic tubes can also send solid capsules using compressed air. For liquids/ gases, any chemically stable liquid or gas can be sent through a pipeline. Short-distance systems exist for sewage, slurry, water and beer, while long-distance networks are used for petroleum and natural gas.

Cable transport is a broad mode where vehicles arc pulled by cables instead of an internal power source. It is most commonly used at steep gradient. Typical solutions include aerial tramway, elevators, escalator and ski lifts; some of these are also categorized as conveyor transport.

Spaceflight is transport out of Earth's atmosphere into outer space by means of a spacecraft. While large amounts of research have gone into technology, it is rarely used except to put satellites into orbit, and conduct scientific experiments. However, man has landed on the moon, and probes have been sent to all the planets of the Solar System.

Suborbital spaceflight is the fastest of the existing and planned transport systems from a place on Earth to a distant other place on Earth. Faster transport could be achieved through part of a Low Earth orbit, or following that trajectory even faster using the propulsion of the rocket to steer it.

Elements

Infrastructure

Infrastructure is the fixed installations that allow a vehicle to operate. It consists of both a way, terminal and facilities for parking and maintenance. For rail, pipeline, road and cable transport, the entire way the vehicle travels must be built up. Air and water craft are able to avoid this, since the airway and seaway do not need to be built up. However, they require fixed infrastructure at terminals.

Terminals such as airports, ports and stations, are locations where passengers and freight can be transferred from one vehicle or mode to another. For passenger transport, terminals are integrating different modes to allow riders to interchange to take advantage of each mode's advantages. For instance, airport rail links connect airports to the city centres and suburbs. The terminals for automobiles are parking lots, while buses and coaches can operates from simple stops. For freight, terminals act as transshipment points, though some cargo is transported directly from the point of production to the point of use.

The financing of infrastructure can either be public or private. Transport is often a natural monopoly and a necessity for the public; roads, and in some countries railways and airports are funded through taxation. New infrastructure projects can involve large spendings, and are often financed through debt. Many infrastructure owners therefore impose usage fees, such as landing fees at airports, or toll plazas on roads. Independent of this, authorities may impose taxes on the purchase or use of vehicles.

Vehicles

A vehicle is any non-living device that is used to move people and goods. Unlike the infrastructure, the vehicle moves along with the cargo and riders. Vehicles that do not operate on land, are usually called crafts. Unless being pulled by a cable or muscle-power, the vehicle must provide its own propulsion; this is most commonly done through a steam engine, combustion engine, electric motor, a jet engine or a rocket, though other means of propulsion also exist. Vehicles also need a system of converting the energy into movement; this is most commonly done through wheels, propellers and pressure.

Vehicles are most commonly staffed by a driver. However, some systems, such as people movers and some rapid transits, are fully automated. For passenger transport, the vehicle must have a compartment for the passengers. Simple vehicles, such as automobiles, bicycles or simple aircraft, may have one of the passengers as a driver.

Operation

Private transport is only subject to the owner of the vehicle, who operates the vehicle themselves. For public transport and freight transport, operations are done through private enterprise or by governments. The infrastructure and vehicles may be owned and operated by the same company, or they may be operated by different entities. Traditionally, many countries have had a national airline and national railway. Since the 1980s, many of these have been privatized. International shipping remains a highly competitive industry with little regulation, but ports can be public owned.

Function

Relocation of travellers and cargo are the most common uses of transport. However, other uses exist, such as the strategic and tactical relocation of armed forces during warfare, or the civilian mobility construction or emergency equipment.

Passenger

Passenger transport, or travel, is divided into public and private transport. Public is scheduled services on fixed routes, while private is vehicles that provide ad hoc services at the riders desire. The latter offers better flexibility, but has lower capacity, and a higher environmental impact. Travel may be as part of daily commuting, for business, leisure or migration.

Short-haul transport is dominated by the automobile and mass transit. The latter consists of buses in rural and small cities, supplemented with commuter rail, trams and rapid transit in larger cities. Long-haul transport involves the use of the automobile, trains, coaches and aircraft, the last of which have become predominantly used for the longest, including intercontinental, travel. Intermodal passenger transport is where a journey is performed through the use of several modes of transport; since all human transport normally starts and ends with walking, all passenger transport can be considered intermodal. Public transport may also involve the intermediate change of vehicle, within or across modes, at a transport hub, such as a bus or railway station.

Taxis and Buses can be found on both ends of Public Transport spectrum, whereas Buses remain the cheaper mode of transport but are not necessarily flexible, and Taxis being very flexible but more expensive. In the middle is Demand responsive transport offering flexibility whilst remaining affordable.

International travel may be restricted for some individuals due to legislation and visa requirements.

Freight

Freight transport, or shipping, is a key in the value chain in manufacturing. With increased specialization and globalization, production is being located further away from consumption, rapidly increasing the demand for transport. While all modes of transport are used for cargo transport, there is high differentiation between the nature of the cargo transport, in which mode is chosen. Logistics refers to the entire process of transferring products from producer to consumer, including storage, transport, transshipment, warehousing, material-handling and packaging, with associated exchange of information. Incoterm deals with the handling of payment and responsibility of risk during transport.

Containerization, with the standardization of ISO containers on all vehicles and at all ports, has revolutionized international and domestic trade, offering huge reduction in transshipment costs. Traditionally, all cargo had to be manually loaded and unloaded into the haul of any ship or car; containerization allows for automated handling and transfer between modes, and the standardized sizes allow for gains in economy of scale in vehicle operation. This has been one of the key driving factors in international trade and globalization since the 1950s.

Bulk transport is common with cargo that can be handled roughly without deterioration; typical examples are ore, coal, cereals and petroleum. Because of the uniformity of the product, mechanical handling can allow enormous quantities to be handled quickly and efficiently. The low value of the cargo combined with high volume also means that economies of scale become essential in transport, and gigantic ships and whole trains are commonly used to transport bulk. Liquid products with sufficient volume may also be transported by pipeline.

Air freight has become more common for products of high value; while less than one percent of world transport by volume is by airline, it amounts to forty percent of the value. Time has become especially important in regards to principles such as postponement and just-in-time within the value chain, resulting in a high willingness to pay for quick delivery of key components or items of high value-to-weight ratio. In addition to mail, common items send by air include electronics and fashion clothing.

History

Humans' first means of transport were walking and swimming. The domestication of animals introduces a new way to lay the burden of transport on more powerful creatures, allowing heavier loads to be hauled, or humans to ride the animals for higher speed and duration. Inventions such as the wheel and sled helped make animal transport more efficient through the introduction of vehicles. Also water transport, including rowed and sailed vessels, dates back to time immemorial, and was the only efficient way to transport large quantities or over large distances prior to the Industrial Revolution.

The first forms of road transport were horses, oxen or even humans carrying goods over dirt tracks that often followed game trails. Paved roads were built by many early civilizations, including Mesopotamia and the Indus Valley Civilization. The Persian and Roman empires built stone-paved roads to allow armies to travel quickly. Deep roadbeds of crushed stone underneath ensured that the roads kept dry. The medieval Caliphate later built tar-paved roads. The first watercraft were canoes cut out from tree trunks. Early water transport was accomplished with ships that were either rowed or used the wind for propulsion, or a combination of the two. The importance of water has led to most cities, that grew up as sites for trading, being located on rivers or at sea, ofter at the intersection of two bodies of water. Until the Industrial Revolution, transport remained slow and costly, and production and consumption were located as close to each other as feasible.

The Industrial Revolution in the 19th century saw a number of inventions fundamentally change transport. With telegraphy, communication became instant and independent of transport. The invention of the steam engine, closely followed by its application in rail transport, made land transport independent of human or animal muscles. Both speed and capacity increased rapidly, allowing specialization through manufacturing being located independent of natural resources. The 19th century also saw the development of the steam ship, that sped up global transport.

The development of the combustion engine and the automobile at the turn into the 20th century, road transport became more viable, allowing the introduction of mechanical private transport. The first highways were constructed during the 19th century with macadam. Later, tarmac and concrete became the dominant paving material. In 1903, the first controllable airplane was invented, and after World War I, it became a fast way to transport people and express goods over long distances.

After World War II, the automobile and airlines took higher shares of transport, reducing rail and water to freight and short-haul passenger. Space flight was launched in the 1950s, with rapid growth until the 1970s, when interest dwindled. In the 1950s, the introduction of containerization gave massive efficiency gains in freight transport, permitting globalization. International air travel became much more accessible in the 1960s, with the commercialization of the jet engine. Along with the growth in automobiles and motorways, this introduced a decline for rail and water transport. After the introduction of the Shinkansen in 1964, high-speed rail in Asia and Europe started taking passengers on long-haul routes from airlines.

Impact

Economic

Transport is a key necessity for specialization—allowing production and consumption of products to occur at different locations. Transport has throughout history been a spur to expansion; better transport allows more trade and a greater spread of people. Economic growth has always been dependent on increasing the capacity and rationality of transport. But the infrastructure and operation of transport has a great impact on the land and is the largest drainer of energy, making transport sustainability a major issue.

Modern society dictates a physical distinction between home and work, forcing people to transport themselves to places of work or study, as well as to temporarily relocate for other daily activities. Passenger transport is also the essence of tourism, a major part of recreational transport. Commerce requires the transport of people to conduct business, either to allow face-to-face communication for important decisions or to move specialists from their regular place of work to sites where they are needed.

Planning

Transport planning allows for high utilization and less impact regarding new infrastructure. Using models of transport forecasting, planners are able to predict future transport patterns. On the operative level, logistics allows owners of cargo to plan transport as part of the supply chain. Transport as a field is studied through transport economics, the backbone for the creation of regulation policy by authorities. Transport engineering, a sub-discipline of civil engineering, and must take into account trip generation, trip distribution, mode choice and route assignment, while the operative level is handles through traffic engineering.

Because of the negative impacts made, transport often becomes the subject of controversy related to choice of mode, as well as increased capacity. Automotive transport can be seen as a tragedy of the commons, where the flexibility and comfort for the individual deteriorate the natural and urban environment for all. Density of development depends on mode of transport, with public transport allowing for better spacial utilization. Good land use keeps common activities close to peoples homes and places higher-density development closer to transport lines and hubs; minimize the need for transport. There are economies of agglomeration. Beyond transportation some land uses are more efficient when clustered. Transportation facilities consume land, and in cities, pavement (devoted to streets and parking) can easily exceed 20 percent of the total land use. An efficient transport system can reduce land waste.

Too much infrastructure and too much smoothing for maximum vehicle throughput means that in many cities there is too much traffic and many—if not all—of the negative impacts that come with it. It is only in recent years that traditional practices have started to be questioned in many places, and as a result of new types of analysis which bring in a much broader range of skills than those traditionally relied on—spanning such areas as environmental impact analysis, public health, sociologists as well as economists who increasingly are questioning the viability of the old mobility solutions. European cities are leading this transition.

Environment

Transport is a major use of energy, and burns most of the world's petroleum. This creates air pollution, including nitrous oxides and particulates, and is a significant contributor to global warming through emission of carbon dioxide, for which transport is the fastest-growing emission sector. By subsector, road transport is the largest contributor to global warming. Environmental regulations in developed countries have reduced the individual vehicles emission; however, this has been offset by an increase in the number of vehicles, and more use of each vehicle. Some pathways to reduce the carbon emissions of road vehicles considerably have been studied. Energy use and emissions vary largely between modes, causing environmentalists to call for a transition from air and road to rail and human-powered transport, and increase transport electrification and energy efficiency.

Other environmental impacts of transport systems include traffic congestion and automobile-oriented urban sprawl, which can consume natural habitat and agricultural lands. By reducing transportation emissions globally, it is predicted that there will be significant positive effects on Earth's air quality, acid rain, smog and climate change.

World Economy

The world economy generally refers to the economy of all of the world's countries. It can be evaluated in various kind of ways. For instance, depending on the model used, the valuation that is arrived at can be represented in a certain currency, such as 2006 US dollars. It is inseparable from the geography and ecology of Earth, and is therefore somewhat of a misnomer, since, while definitions and representations of the "world economy" vary widely, they must at a minimum exclude any consideration of resources or value based outside of the Earth.

For example, while attempts could be made to calculate the value of currently unexploited mining opportunities in unclaimed territory in Antarctica, the same opportunities on Mars would not be considered a part of the world economy—even if currently exploited in some way—and could be considered of latent value only in the same way as uncreated intellectual property, such as a previously unconceived invention.

Beyond the minimum standard of concerning value in production, use, and exchange on the planet Earth, definitions, representations, models, and valuations of the world economy vary widely.

It is common to limit questions of the world economy exclusively to human economic activity, and the world economy is typically judged in monetary terms, even in cases in which there is no efficient market to help valuate certain goods or services, or in cases in which a lack of independent research or government cooperation makes establishing figures difficult.

Typical examples are illegal drugs and other black market goods, which by any standard are a part of the world economy, but for which there is by definition no legal market of any kind.

However, even in cases in which there is a clear and efficient market to establish a monetary value, economists do not typically use the current or official exchange rate to translate the monetary units of this market into a single unit for the world economy, since exchange rates typically do not closely reflect worldwide value, for example in cases where the volume or price of transactions is closely regulated by the government.

Rather, market valuations in a local currency are typically translated to a single monetary unit using the idea of purchasing power. This is the method used below, which is used for estimating worldwide economic activity in terms of real US dollars.

However, the world economy can be evaluated and expressed in many more ways. It is unclear, for example, how many of the world's 6.8 billion people have most of their economic activity reflected in these valuations.

Economy - Overview

2007–2008

Global output (gross world product) (GWP) rose by 3.2% in 2008, led by China (9%, equal to 21% of global growth), the US (1.1%, or 12% of growth), the European Union (0.9%, for a 10.5% share of growth) and India (7.3%, equal to 5.6% of the total rise). The 12 largest economies (the US, Japan, China, Germany, France, the United Kingdom, Italy, Russia, Spain, Brazil, Canada and India) contributed just over half of all economic growth in 2008.

Growth results in the wealthy, or "advanced" economies, slowed by two-thirds, from 2.7% in 2007 to just 0.9% in 2008. Emerging Asia slowed from 9.8% to 6.8%; Emerging Europe from 5.4% to 2.9%; the Commonwealth of Independent States from 8.6% to 5.5%; the (non-OECD) Western Hemisphere from 5.7% to 4.2%; the Middle East from 6.3% to 5.9%; and Africa from 6.2% to 5.2%.

Externally, the nation-state, as a bedrock economic-political institution, is steadily losing control over international flows of people, goods, funds, and technology. Central governments are losing decision making powers and enhancing their international collective power thanks to strong economic bodies of which they democratically chose to become part, notably the EU. The introduction of the euro as the common currency of much of Western Europe in January 1999, while paving the way for an integrated economic powerhouse, poses economic risks because of varying levels of income and cultural and political differences among the participating nations.

Internally, the central government often finds its control over resources slipping as separatist regional movements-typically based on ethnicity-gain momentum, e.g., in many of the successor states of the former Soviet Union, in the former Yugoslavia, in India, in Iraq, and in Indonesia.

Statistical Indicators

Economy:

- GDP (GWP) (gross world product): (purchasing power parity exchange rates)-$59.38 trillion 2005 est.), $51.48 trillion (2004), $23 trillion (2002)

- GDP (GWP) (gross world product):'" (market exchange rates)-$60.69 trillion (2008)
- GDP-real growth rate: 3.2% (2008), 3.1% p.a. (2000-07), 2.4% p.a. (1990-99), 3.1% p.a. (1980-89)
- GDP-per capita: purchasing power parity-$9,300 (2005 est.), $8,200 (92) (2003), $7,900 (2002)
- GDP-composition by sector: agriculture: 4%; industry: 32%; services: 64% (2004 est.)
- Inflation rate (consumer prices): developed countries 1% to 4% typically; developing countries 5% to 60% typically; national inflation rates vary widely in individual cases, from declining prices in Japan to hyperinflation in several Third World countries (2003)
- Derivatives outstanding notional amount: $273 trillion (end of June 2004), $84 trillion (end-June 1998)
- Global debt issuance: $5.187 trillion (2004), $4.938 trillion (2003), $3.938 trillion (2002) (Thomson Financial League Tables)
- Global equity issuance: $505 billion (2004), $388 billion (2003), $319 billion (2002) (Thomson Financial League Tables).

Employment:

- Unemployment rate: 30% combined unemployment and underemployment in many non-industrialized countries; developed countries typically 4%-12% unemployment.

Industries:

- Industrial production growth rate: 3% (2002 est.)

Energy:

- Yearly electricity-production: 15,850,000 GWh (2003 est.), 14,850,000 GWh (2001 est.)
- Yearly electricity-consumption: 14,280,000 GWh (2003 est.), 13,930,000 GWh (2001 est.)
- Oil-production: 79.65 million bbl/day (2003 est.), 75.46 million barrel/day (12,000,000 m^3/d) (2001)
- Oil-consumption: 80.1 million bbl/day (2003 est.), 76.21 million barrel/day (12,120,000 m^3/d) (2001)

- Oil-proved reserves: 1.025 trillion barrel (163 km³) (2001 est.)
- Natural gas-production: 2,569 km³ (2001 est.)
- Natural gas-consumption: 2,556 km³ (2001 est.)
- Natural gas-proved reserves: 161,200 km³ (1 January 2002)

Cross-border

- Yearly exports: $1.1 trillion (f.o.b., 2002 est.)
- Exports-commodities: the whole range of industrial and agricultural goods and services
- Exports-partners: US 17.4%, Germany 7.6%, UK 5.4%, France 5.1%, Japan 4.8%, China 4% (2002)
- Yearly imports: $2.5 trillion (f.o.b., 2002 est.)
- Imports-commodities: the whole range of industrial and agricultural goods and services
- Imports-partners: US 11.2%, Germany 9.2%, China 7%, Japan 6.8%, France 4.7%, UK 4% (2002)
- Debt-external: $2 trillion for less developed countries (2002 est.).

Gift economy:

- Yearly economic aid-recipient: Official Development Assistance (ODA) $50 billion

Communications

Telephones-main lines in use: 843,923,500 (2007), 4,263,367,600 (2008)

- Telephones-mobile cellular: 3,300,000,000 (Nov. 2007)
- Internet Service Providers (ISPs): 10,350 (2000 est.)
- Internet users: 1,311,050,595 (January 18, 2008 est.), 1,091,730,861 (December 30, 2006 est.), 604,111,719 (2002 est.)

Transport

Transportation infrastructure worldwide includes:

- Airports
  - o Total: 49,973 (2004)

- Roadways (in kilometres)
  - o Total: 32,345,165 km
  - o Paved: 19,403,061 km
  - o Unpaved: 12,942,104 km (2002)
- Railways
  - o Total: 1,122,650 km includes about 190,000 to 195,000 km of electrified routes of which 147,760 km are in Europe, 24,509 km in the Far East, 11,050 km in Africa, 4,223 km in South America, and 4,160 km in North America.

Military:

- Military expenditures-dollar figure: aggregate real expenditure on arms worldwide in 1999 remained at approximately the 1998 level, about $750 billion, about 1/2 of which was the United States (1999)
- Military expenditures-percent of GDP: roughly 2% of gross world product (1999).

Trade Route

A trade route is a logistical network identified as a series of pathways and stoppages used for the commercial transport of cargo. Allowing goods to reach distant markets, a single trade route contains long distance arteries which may further be connected to several smaller networks of commercial and non commercial transportation.

Historically, the period from 1532 BCE–1 CE saw the Western Asian, Mediterranean, Chinese and Indian societies develop major transportation networks for trade. Europe's early trading routes included the Amber Road, which served as a dependable network for long distance trade. Maritime trade along the Spice route became prominent during the Middle Ages; nations resorted to military means for control of this influential route. During the Middle Ages organizations such as the Hanseatic League, aimed at protecting interests of the merchants and trade, also became increasingly prominent.

With the advent of modern times, commercial activity shifted from the major trade routes of the Old World to newer routes between modern nation states. This activity was sometimes carried out without traditional protection of trade and under international free trade agreements, which allowed commercial goods to cross borders with relaxed restrictions. Innovative transportation of the modern times includes pipeline transport, and the relatively well known trade using rail routes, automobiles and cargo airlines.

Development of Early Routes

Early Development

The period extending from the middle of the 2nd millennium BCE to the beginning of the Common Era saw the Western Asian, Mediterranean, Chinese and Indian societies develop major transportation networks for trade.

One of the vital instruments which facilitated long distance trade was portage and the domestication of beasts of burden. Organized caravans, visible by the 2nd millennium BCE, could carry goods across a large distance as fodder was mostly available along the way. The domestication of camels allowed Arabian nomads to control the long distance trade in spices and silk from the Far East to the Arabian Peninsula. However, caravans were useful in long-distance trade largely for carrying luxury goods, the transportation of cheaper goods across large distances was not profitable for caravan operators. With productive developments in iron and bronze technologies, newer trade routes-dispensing innovations of civilizations-began to rise slowly.

Maritime Trade

Visible maritime trade between civilizations can be traced back to at least two millennia. Navigation was known in Sumer between the 4th and the 3rd millennium BCE, and was probably known by the Indians and the Chinese people before the Sumerians. The Egyptians had trade routes through the Red sea, importing spices from the "Land of Punt" (East Africa) and from Arabia.

Maritime trade began with safer coastal trade and evolved with the manipulation of the monsoon winds, soon resulting in trade crossing boundaries such as the Arabian Sea and the Bay of Bengal. South Asia had multiple maritime trade routes which connected it to Southeast Asia, thereby making the control of one route resulting in maritime monopoly difficult. Indian connections to various Southeast Asian states buffered it from blockages on other routes. By making use of the maritime trade routes, bulk commodity trade became possible for the Romans in the 2nd century BCE. A Roman trading vessel could span the Mediterranean in a month at one-sixtieth the cost of over-land routes.

Visible Trade Routes

The peninsula of Anatolia lay on the commercial land routes to Europe from Asia as well as the sea route from the Mediterranean to the Black Sea. Records from the 19th century BCE attest to the existence of an Assyrian merchant colony at Kanesh in Cappadocia (now in modern Turkey). Trading networks of the Old World included the Grand Trunk Road of India and the Incense Road of Arabia.

A transportation network consisting of hard-surfaced highways, using concrete made from volcanic ash and lime, was built by the Romans as early as 312 BCE, during the times of the Censor Appius Claudius Caecus. Parts of the Mediterranean world, Roman Britain, Tigris-Euphrates river system and North Africa fell under the reach of this network at some point of their history.

According to Robert Allen Denemark (2000):

> *"The spread of urban trading networks, and their extension along the Persian Gulf and eastern Mediterranean, created a complex molecular structure of regional foci so that as well as the zonation of core and periphery (originally created around Mesopotamia) there was a series of interacting civilizations: Mesopotamia, Egypt, the Indus Valley; then also Syria, central Anatolia (Hittites) and the Aegean (Minoans and Mycenaeans). Beyond this was a margin which included not only temperate areas such as Europe, but the dry steppe corridor of central Asia. This was truly a world system, even though it occupied only a restricted portion of the western Old World. Whilst each civilization emphasized its ideological autonomy, all were identifiably part of a common world of interacting components."*

These routes-spreading religion, trade and technology-have historically been vital to the growth of urban civilization. The extent of development of cities, and the level of their integration into a larger world system, has often been attributed to their position in various active transport networks.

Historic Trade Routes

Incense Route:

The Incense Route served as a channel for trading of Indian, Arabian and East Asian goods. The incense trade flourished from South Arabia to the Mediterranean between roughly the 3rd century BCE to the 2nd century CE. This trade was crucial to the economy of Yemen and the frankincense and myrrh trees were seen as a source of wealth by the its rulers.

Ptolemy II Philadelphus, emperor of Ptolemaic Egypt, may have forged an alliance with the Lihyanites in order to secure the incense route at Dedan, thereby rerouting the incense trade from Dedan to the coast along the Red Sea to Egypt. I. E. S. Edwards connects the Syro-Ephraimite War to the desire of the Israelites and the Aramaeans to control the northern end of the Incense route, which ran up from Southern Arabia and could be tapped by commanding Transjordan.

Gerrha-inhabited by Chaldean exiles from Babylon-controlled the Incense trade routes across Arabia to the Mediterranean and exercised control over the trading of aromatics to Babylon in the 1st century BC. The Nabateans exercised control over the routes along the Incense Route, and their hold was challenged-without success-by Antigonus Cyclops, emperor of Syria and Palestine. The Nabatean control over trade further increased and spread in many directions.

The replacement of Greece by the Roman empire as the administrator of the Mediterranean basin led to the resumption of direct trade with the East and the elimination of the taxes extracted previously by the middlemen of the south. According to Milo Kearney (2003) "The South Arabs in protest took to pirate attacks over the Roman ships in the Gulf of Aden. In response, the Romans destroyed Aden and favoured the Western Abyssinian coast of the Red Sea." Indian ships sailed to Egypt as the maritime routes of Southern Asia were not under the control of a single power.

Pre-Columbian Trade

Some similarities between the Mesoamerican and the Andean cultures suggest that the two regions became a part of a wider world system, as a result of trade, by the 1st millennium BCE. The current academic view is that the flow of goods across the Andean slopes was controlled by institutions distributing locations to local groups, who were then free to access them for trading. This trade across the Andean slopes-described sometimes as "vertical trade"-may have overshadowed the long distance trade between the people of the Andes and the neighbouring forests. The Callawaya herbalists traded in tropical plants between 6th and the 10th centuries, while copper was dealt by specialized merchants in the Peruvian valley of Chincha. Long distance trade may have seen local elites resorting to struggle in order for manipulation and control.

Prior to the Inca dominance, specialized long distance merchants provided the highlanders with goods such as gold nuggets, copper hatches, cocoa, salt etc. for redistribution among the locals, and were key players in the politics of the region. Hatchet shaped copper currency was produced by the Peruvian people, in order to obtain valuables from pre Columbian Ecuador. A maritime exchange system stretched from the west coast of Mexico to southernmost Peru, trading mostly in Spondylus, which represented rain, fertility and was considered the principal food of the gods by the people of the Inca empire. Spondylus was used in elite rituals and the effective redistribution of it had political effect in the Andes during the pre-Hispanic times.

Predominantly Overland Routes

Silk Route

The Silk road was one of the first trade routes to join the Eastern and the Western worlds. According to Vadime Elisseeff (2000):

> *"Along the Silk Roads, technology travelled, ideas were exchanged, and friendship and understanding between East and West were experienced for the first time on*

a large scale. Easterners were exposed to Western ideas and life-styles, and Westerners too, learned about Eastern culture and its spirituality-oriented cosmology. Buddhism as an Eastern religion received international attention through the Silk Roads."

Cultural interactions patronized often by powerful emperors, such as Kanishka, led to development of art due to introduction of a rich variety of influences. Buddhist missions thrived along the Silk Roads, partly due to the conducive intermixing of trade and cultural values, which created a series of safe stoppages for both the pilgrims and the traders. Among the frequented routes of the Silk Route was the Burmese route extending from Bhamo, which served as a path for Marco Polo's visit to Yunnan and Indian Buddhist missions to Canton in order to establish Buddhist monasteries. This route-often under the presence of hostile tribes-also finds mention in the works of Rashid-al-Din Hamadani.

Grand Trunk Road

The Grand Trunk Road-connecting Calcutta in India to Peshawar in Pakistan-has existed for over two and a half millennia. One of the important trade routes of the world, this road has been a strategic artery with fortresses, halting posts, wells, post offices, milestones and other facilities. Part of this road through Pakistan also coincided with the Silk Road.

This highway has been associated with emperors Chandragupta Maurya and Sher Shah Suri, the latter became synonymous with this route due to his role in ensuring the safety of the travellers and the upkeep of the road. Emperor Sher Shah widened and realigned the road to other routes, and provided approximately 1700 roadside inns through his empire. These inns provided free food and lodgings to the travellers regardless of their status.

The British occupation of this road was of special significance for the British Raj in India. Bridges, pathways and newer inns were constructed by the British for the first thirty seven years of their reign since the occupation of Punjab in 1849. The British followed roughly the same alignment as the old routes, and at some places the newer routes ran parallel to the older routes.

Vadime Elisseeff (2000) comments on the Grand Trunk Road:

> *"Along this road marched not only the mighty armies of conquerors, but also the caravans of traders, scholars, artists, and common folk. Together with people, moved ideas, languages, customs, and cultures, not just in one, but in both directions. At different meeting places-permanent as well as temporary-people of different origins and from different cultural backgrounds, professing different faiths and creeds, eating different foods, wearing different clothes, and speaking different languages and dialects would meet one another peacefully. They would understand one another's food, dress, manner, and etiquette, and even borrow words, phrases, idioms and, at times, whole languages from others."*

Amber Road

The Amber Road was a European trade route associated with the trade and transport of amber. Amber satisfied the criteria for long distance trade as it was light in weight and was in high demand for ornamental purposes around the Mediterranean. Before the establishment of Roman control over areas such as Pannonia, the Amber Road was virtually the only route available for long distance trade.

Towns along the Amber Road began to rise steadily during the 1st century CE, despite the troop movements under Titus Flavius Vespasianus and his son Titus Flavius Domitianus. Under the reign of Tiberius Caesar Augustus, the Amber Road was straightened and paved according to the prevailing urban standards. Roman towns began to appear along the road, initially founded near the site of Celtic oppida.

The 3rd century saw the Danube river become the principal artery of trade, eclipsing the Amber Road and other commercial routes. The redirection of investment to the Danubian forts saw the towns along the Amber Road growing slowly, though yet retaining their prosperity.

The prolonged struggle between the Romans and the barbarians further left its mark on the towns along the Amber Road.

Via Maris

Via Maris, literally Latin for "the way of the sea," was an ancient highway used by the Romans and the Crusaders. The states controlling the Via Maris were in a position to grant access for trade to their own citizens and collect tolls from the outsiders to maintain the trade route. The name Via Maris is a Latin translation of a Hebrew phrase related to Isaiah. Due to the Biblical significance of this ancient route, many attempts to find its present day location have been made by Christian pilgrims. 13th century traveller and pilgrim Burchard of Mount Zion refers to the Via Maris route as a way leading along the shore of the Sea of Galilee.

Trans Saharan Trade

Early Muslim writings confirm that the people of West Africa operated a sophisticated network of trade, usually under the authority of a monarch who levied taxes and provided bureaucratic and military support to his kingdom. Sophisticated mechanisms for the economic and political development of the involved African areas were in place before Islam further strengthened trade, towns and government in western Africa. The capital, court and trade of the region find mention in the works of scholar Abu 'Ubayd 'Abd Allah al-Bakri; the mainstay of the trans Saharan trade was gold and salt.

The powerful Saharan tribes, Berber in origin and later adapting to Muslim and Arab cultures, controlled the channels to western Africa by making efficient use of horse-drawn vehicles and pack animals. The Songhai engaged in a struggle against the Sadi dynasty of Morocco over the control of the trans Saharan trade, resulting in damage on both sides and a weak Moroccan victory, further strengthening the uninvolved Saharan tribes. Struggles and disturbances continued till the 14th century, by which the Mande merchants were trading with the Hausa, between Lake Chad and the Niger. Newer trade routes developed following extension of trade.

Predominantly Maritime Routes

Roman-India Routes

The Ptolemaic dynasty had initiated Greco-Roman maritime trade contact with India using the Red Sea ports. The Roman historian Strabo mentions a vast increase in trade following the Roman annexation of Egypt, indicating that monsoon was known and manipulated for trade in his time. By the time of Augustus up to 120 ships were setting sail every year from Myos Hormos to India, trading in a diverse variety of goods. Arsinoe, Berenice Troglodytica and Myos Hormos were the principal Roman ports involved in this maritime trading network, while the Indian ports included Barbaricum, Barygaza, Muziris and Arikamedu.

The Indians were present in Alexandria and the Christian and Jewish settlers from Rome continued to live in India long after the fall of the Roman empire, which resulted in Rome's loss of the Red Sea ports, previously used to secure trade with India by the Greco-Roman world since the time of the Ptolemaic dynasty.

Hanseatic Trade

Shortly before the 12th century the Germans played a relatively modest role in the north European trade. However, this was to change with the development of Hanseatic trade, as a result of which German traders became prominent in the Baltic and the North Sea regions.

Following the death of Eric VI of Denmark, German forces attacked and sacked Denmark, bringing with them artisans and merchants under the new administration which controlled the Hansa regions. During the third quarter of the 14th century the Hanseatic trade faced two major difficulties: economic conflict with the Flanders and hostilities with Denmark. These events led to the formation of an organized association of Hanseatic towns, which replaced the earlier union of German merchants. This new Hansa of the towns-aimed at protecting interests of the merchants and trade-became prominent for the next hundred and fifty years.

Philippe Dollinger associates the downfall of the Hansa to a new alliance between Lübeck, Hamburg and Bremen, which outshadowed the older institution. He further sets the date of dissolution of the Hansa at 1630 and concludes that the Hansa was almost entirely forgotten by the end of the 18th century. Scholar Georg Friedrich Sartorius published the first monograph regarding the community in the early years of the 19th century.

Maritime Republics' Mediterranean Trade

The economic growth of Europe around the year 1000, together with the lack of safety on the mainland trading routes, eased the development of major commercial routes along the coast of the Mediterranean. The growing independence of some coastal cities gave them a leading role in this comerce: Maritime Republics (Italian "Repubbliche Marinare") of Venice, Genoa, Amalfi, Pisa and Republic of Ragusa developed their own "empires" in the Mediterranean shores. From the 8th until the 15th century, they held the monopoly of European trade with the Middle East. The silk and spice trade, involving spices, incense, herbs, drugs and opium, made these Mediterranean city-states phenomenally rich. Spices were among the most expensive and demanded products of the Middle Ages. They were all imported from Asia and Africa. Muslim traders—mainly descendants of Arab sailors from Yemen and Oman—dominated maritime routes throughout the Indian Ocean, tapping source regions in the Far East and shipping for trading emporiums in India, westward to Ormus in Persian Gulf and Jeddah in the Red Sea. From there, overland routes led to the Mediterranean coasts. Venetian merchants distributed then the goods through Europe until the rise of the Ottoman Empire, that eventually led to the fall of Constantinople in 1453, barring Europeans from important combined-land-sea routes.

Spice Route

As trade between India and the Greco-Roman world increased spices became the main import from India to the Western world, bypassing silk and other commodities. The Indian commercial connection with South East Asia proved vital to the merchants of Arabia and Persia during the 7th century and the 8th century.

The Abbasids used Alexandria, Damietta, Aden and Siraf as entry ports to India and China. Merchants arriving from India in the port city of Aden paid tribute in form of musk, camphor, ambergris and sandalwood to Ibn Ziyad, the sultan of Yemen. Moluccan products shipped across the ports of Arabia to the Near East passed through the ports of India and Sri Lanka. Indian exports of spices find mention in the works of Ibn Khurdadhbeh (850), al-Ghafiqi (1150 AD), Ishak bin Imaran

(907) and Al Kalkashandi (14th century). After reaching either the Indian or the Sri Lankan ports, spices were sometimes shipped to East Africa, where they were used for many purposes, including burial rites.

On the orders of Manuel I of Portugal, four vessels under the command of navigator Vasco da Gama rounded the Cape of Good Hope, continuing to the eastern coast of Africa to Malindi to sail across the Indian Ocean to Calicut. The wealth of the Indies was now open for the Europeans to explore; the Portuguese Empire was one of the early European empires to grow from spice trade.

Modern Routes

The modern times saw development of newer means of transport and often controversial free trade agreements, which altered the political and logistical approach prevalent during the Middle Ages. Newer means of transport led to the establishment of new routes, and countries opened up borders to allow trade in mutually agreed goods as per the prevailing free trade agreement. Some old trading route were reopened during the modern times, although in different political and logistical scenarios. The entry of harmful foreign pollutants by the way of trade routes has been a cause of alarm during the modern times. A conservative estimate stresses that future damages from harmful animal and plant diseases may be as high as 134 billion US dollars in the absence of effective measures to prevent the introduction of unwanted pests through various trade routes.

Wagonway Routes

Networks, like the Santa Fe Trail and the Oregon Trail, became prominent in the United States with wagon trains gaining popularity as a mode of long distance overland transportation for both people and goods. The Oregon-California routes were highly organized with planned rendezvous locations and essential supplies. The settlers in the United States used these wagon trains-sometimes made up of 100 of more Conestoga wagons-for westward emigration during the 18th and the 19th centuries. Among the challenges faced by the wagon route operators were crossing rivers, mountains and hostile Native

Americans. Preparations were also made according to the weather and protection of trade and travellers was ensured by a few guards on horseback. Wagon freighting was also essential to American growth until it was replaced by the railroad and the truck.

Railway Routes

The 1844 Railway act of England compelled at least one train to a station every day with the third class fares priced at a penny a mile. Trade benefited as the workers and the lower classes had the ability to travel to other towns frequently. Suburban communities began to develop and towns began to spread outwards. The British constructed a vast railway network in India, but it was considered to serve a strategic purpose in addition to the commercial purpose. The efficient use of rail routes helped in the unification of the United States of America.

The modern times saw nations struggle for the control of rail routes: The Trans-Siberian Railway was intended to be used by the Russian government for control of Manchuria and later China; the German forces wanted to establish Berlin-Baghdad Railway in order to influence the Near East; and the Austrian government planned a route from Vienna to Salonika for control of the Balkans. According to the Encyclopædia Britannica (2002):

Railroads reached their maturity in the early 20th century, as trains carried the bulk of land freight and passenger traffic in the industrialized countries of the world. By the mid-20th century, however, they had lost their preeminent position. The private automobile had replaced the railroad for short passenger trips, while the airplane had usurped it for long-distance travel, especially in the United States. Railroads remained effective, however, for transporting people in high-volume situations, such as commuting between the centres of large cities and their suburbs, and medium-distance travel of less than about 300 miles between urban centres. Although railroads have lost much of the general-freight-carrying business to semi-trailer trucks, they remain the best means of transporting large volumes of such bulk commodities as coal, grain, chemicals, and ore over long distances. The development of containerization has made the railroads more effective in handling finished merchandise at relatively high speeds. In addition, the introduction of piggyback flatcars, in which truck trailers are transported long distances on specially-designed cars, has allowed railroads to regain some of the business lost to trucking.

Modern Road Networks

The advent motor vehicles created a demand for better use of highways. Roads evolved into two way roads, expressways, freeways and tollways during the modern times. Existing roads were developed and highways were designed according to intended use. Trucks came into widespread use in the Western World during World War I, and quickly gained reputation as a means of long distance transportation of goods. Modern highways, such as the Trans-Canada Highway, Highway 1 (Australia) and Pan-American Highway allowed transport of goods and services across great distances. Automobiles continue to play a crucial role in the economies of the Industrialized countries, resulting in rise of businesses such as motor freight operation and truck transportation.

The emission rate for cars using highways has been on a decline between 1975 and 1995 due to regulations and the introduction of unleaded petrol. This trend is especially notable since there has been a growth in vehicles and vehicle miles travelled by automobiles using these highways.

Modern Maritime Routes

A consistent shift from land based trade to sea based trade has been recorded since the last three millennia. The strategic advantages of port cities as trading centres are many: they are both less dependent on vital connections and less vulnerable to blockages. Oceanic ports can help forge trading relationships with other parts of the world easily.

Modern maritime trade routes-sometimes in the form of artificial canals like the Suez Canal-had visible impact on the economic and political standing of nations. The opening of the Suez Canal altered British interactions with the colonies of the British Empire as the dynamics of transportation, trade and communication had now changed drastically. Other waterways, like the Panama Canal played an important role in the histories of many nations. Inland water transportation remained significantly important even as the advent of railroads and automobiles resulted in a steady decline of canals. Inland water transport is still used for the transportation of bulk commodities e.g. grains, coal, and ore.

Waterway commerce was historically important to Europe, particularly to Russia. According to the Encyclopædia Britannica (2002): "Russia has been a significant beneficiary. Not only have inland waterways opened vast areas of its interior to development, but Moscow-linked to the White, Baltic, Black, Caspian, and Azov seas by canals and rivers-has become a major inland port."

Oil spills are recorded both in case of maritime routes and pipeline routes to the main refineries. Oil spills, amounting to as much as 7.56 billion liters of oil entering the oceans every year, occur due to damaged equipment or human error.

Free Trade Areas

Historically, governments followed a policy of protection of trade. International Free Trade became visible in 1860 with the Anglo-French commercial treaty and the sentiment further gained momentum during the post World War II era. According to The Columbia Encyclopedia, Sixth Edition:

"After World War II, strong sentiment developed throughout the world against protection and high tariffs and in favour of freer trade. The results were new organizations and agreements on international trade such as the General Agreement on Tariffs and Trade (1948), the Benelux Economic Union (1948), the European Economic Community (Common Market, 1957), the European Free Trade Association (1959), Mercosur (the Southern Cone Common Market, 1991), and the World Trade Organization (1995). In 1993, the North American Free Trade Agreement (NAFTA) was approved by the governments of Canada, Mexico, and the United States. In the early 1990s, the nations of the European Union (the successor organization to the Common Market) undertook to remove all barriers to the free movement of trade and employment across their mutual borders."

On May 2004, the United States of America signed the American Free trade Agreement with five Central American nations.

Air Routes

Air transport has become an indispensable part of the modern society. People having been using air transport both for long and middle distances, with the average route length of long distances being 720 kilometers in Europe and 1220 kilometers in the US. This enormous industry annually carries 1600 million passengers worldwide, and covers a 15 million kilometre network with an annual turnover of 260 billion dollars.

The national, international and global economies are linked to this mode of transportation, making it vital to many other industries. Newer trends of liberalization of trade have further led to establishment of routes among nations bound by agreements. One such example is the American Open Skies policy, which led to greater openness in many international markets, but some international restrictions have survived even during the present times.

Express delivery through international cargo airlines touched US $ 20 billion in 1998 and, according to the World Trade Organization, is expected to triple in 2015. In 1998, 50 pure cargo service companies operated internationally.

Pipeline Networks

The economic importance of pipeline transport-responsible for a high percentage of oil and natural gas transportation-is often undermined by the general public due to the lack of visibility of this mode. Generally held to be safer and more economical and reliable than the other modes of transport, this mode has many advantages over rival modes, such as trucks and railways. Examples of modern pipeline transport include Alashankou-Dushanzi Crude Oil Pipeline and Iran-Armenia Natural Gas Pipeline. International pipeline transport projects, like the Baku-Tbilisi-Ceyhan pipeline, presently connect modern nation states-in this case Azerbaijan, Georgia and Turkey-through pipeline networks.

In some select cases, pipelines can even transport solids, such as coal and other minerals, over long distances; short distance transportation of goods such as grain, cement, concrete, solid wastes, pulp etc. is also feasible.

4

Communication—By Wire and Wireless

Communications in the United States

The primary regulator of communications in the United States is the Federal Communications Commission. It closely regulates all of the industries mentioned below with the exception of newspapers and the Internet service provider industry.

History

Press

Newspapers declined in their influence and penetration into American households in the late 20th century. Most newspapers are local, having little circulation outside their particular metropolitan area. The closest thing to a national paper the U.S. has is USA Today. Other influential dailies include the New York Times, the Washington Post and the Wall Street Journal which are sold in most U.S. cities. The Times has a moderate-left stance, while the Journal is moderate-right and is strongly pro-business.

The largest newspapers (by circulation) in the United States are USA Today, the Wall Street Journal, the New York Times and the Los Angeles Times.

Mail

The legal monopoly of the government-owned United States Postal Service has narrowed during the 20th and 21st centuries, although the USPS, through whose hands passes 40% of the world's mail, still delivers in four days more mail than is delivered by DHL Express, FedEx, and the United Parcel Service in one year.

Landline Telephone

Telephones-main lines in use: 163.2 million (2007)

Telephones-mobile cellular: 255 million (2007))

General assessment: A large, technologically advanced, multipurpose communications system.

Domestic: A large system of fiber-optic cable, microwave radio relay, coaxial cable, and domestic satellites carries every form of telephone traffic; a rapidly growing cellular system carries mobile telephone traffic throughout the country.

International: *Country code*-1; 24 ocean cable systems in use; satellite earth stations-61 Intelsat (45 Atlantic Ocean and 16 Pacific Ocean), 5 Intersputnik (Atlantic Ocean region), and 4 Inmarsat (Pacific and Atlantic Ocean regions) (2000).

- Most of the American telephone system was formerly operated by a single monopoly, AT&T, which was split up in 1984 into a long distance telephone company and several local "Baby Bells."
- Landline telephone service continues to be divided between incumbent local exchange carriers and several competing long distance companies. Most states have several competing cellular phone networks. As of 2005, some of the Baby Bells are beginning to merge with long distance phone companies. A small number of consumers are currently experimenting with Voice over Internet Protocol phone service.
- Most local loop service to homes is provided through old-fashioned copper wire, although many of the Baby Bells are beginning to upgrade the so-called "last mile" to fiber optic.
- Early in the 21st century the number of wire lines in use stopped growing and in some markets began to decline.

Radio

Radio broadcast stations: AM: 4,789; FM commercial stations: 6,231; FM educational stations: 2,672; FM translators & boosters: 3,995; low-power FM stations: 675 (as of December 31, 2005, according to the Federal Communications Commission)

Most broadcast stations are controlled by large media conglomerates like Clear Channel Communications. There are also many small independent local stations. National Public Radio (NPR) is the public radio network.

Radios: 575 million (1997)

Television

Television broadcast stations: 9,024 (of which 1,750 are full-power TV stations; 592 are class-A TV stations; 4,537 are TV translators; and 2,145 are other low-power TV stations) (as of December 31, 2005, according to the Federal Communications Commission); in addition, there are about 12,000 cable TV systems.

Most local commercial television stations are owned-and-operated by or affiliated with the large national broadcast networks such as ABC, CBS, Fox, NBC, and The CW. Some television networks are aimed at ethnic minorities, including Spanish language networks Univision and Telemundo. Public Broadcasting Service (PBS), is the public broadcasting network, with over 300 non-profit affiliated stations across the United States. Besides the large broadcast networks (which are free for anyone with a TV and an antenna), there are also many networks available only with a subscription to cable or satellite television, like CNN.

Televisions: 219 million (1997).

Cellular Communication

The major cellphone companies in the U.S. are AT&T Mobility, Verizon Wireless, Sprint Nextel, and T-Mobile

Internet

The predecessor of the Internet was a project created by the United States Department of Defence in the late 1960s. Called the ARPANET (Advanced Research Projects Agency Network), it was first developed by DARPA (Defence Advanced

Research Projects Agency; at the time simply named ARPA). The Internet, which grew out of this initial project, first gained a public face in the 1990s.

By 2004, three quarters of Americans had Internet access at home. Access is available in a variety of forms.

Common Access Technologies

Access technologies generally use a modem, which converts digital data to analog for transmission over a particular analog network (ex. the telephone or cable networks).

Dial-up

This technology dials into the network through an existing phone line, creating a semi-permanent link to the Internet. Operating on a single channel, it monopolizes the phone line and is the slowest method of accessing the Internet. Dial-up is often the only form of Internet access available in rural areas as it requires no infrastructure, other than the already existing telephone network, to connect to the Internet. Typically, dial-up connections do not exceed a speed of 56 kbit/s, as they are primarily made via a 56k modem.

Broadband

This term includes a broad range of technologies, all of which provide high data rate access to the Internet. Broadband provides a continuous connection; there is no dial-up/in process required and it does not "hog" phone lines. In 2008, the United States Federal Communications Commission (FCC) defined broadband as any connection above 768 kbit/s.

DSL (ADSL/SDSL)

DSL (Digital Subscriber Line) provides a connection to the Internet through the telephone network. Unlike dial-up, DSL can operate using a single phone line without preventing normal use of the telephone line for phone calls. DSL uses the high frequencies, while the low frequencies of the line are left free for regular telephone communication.

WiFi

Wi-Fi is the standard method to connect a high-speed local area network via wireless transmitter/receiver. WiFi is convenient for mobile Internet users and can bring service to areas where wiring would be costly. WiFi service range is short, and penetration through building walls is limited.

WiMax

WiMax (Worldwide Interoperability for Microwave Access) works similarly to WiFi in that it transmits information via airwaves, but it handles network traffic more efficiently. This technology penetrates building walls much more effectively and can be used across larger distances than WiFi.

Cable Modem

Cable Modem transmits data via airwaves on the cable television infrastructure. Although cables have low interference, comparably high speeds, and allow television use, the cost of bringing service into an area (trenching cable) can be very high.

Satellite

Data is transmitted via satellite dishes, which send and receive data to and from satellites in orbit in space. There are many factors that affect the reception of the Internet using this technology, including wind, rain, and trees. However, it is ideal for those living in rural areas where other internet access technologies have not been extended.

Cellphones

Cellular broadband is wireless high-speed internet access through a portable telephone or cellular device. Information is transmitted through cellular towers.

Fibre-optic

Fiber optic wires convert electrical signals carrying data into light, and send this light through tiny transparent glass fibers. This method is extremely fast and has little interference. It is more expensive than other methods to deploy. Its speed is dependent on how close the fibers are to one's computer, the amount of bandwidth available, and how the service is configured.

Power-line Internet

Power-line Internet technology uses power lines to send and receive radio signals. Because of the extensive power line infrastructure already in place, this technology would allow people in rural and low population areas to access the Internet with little cost in terms of equipment, cables or wires.

T-lines

In T-line internet access data is carried over a fiber optic line or copper line. They are fast but highly regulated and generally intended for business use.

Access Policy

The internet became a public entity in the 1990s with the advent of the World Wide Web project. As a result, Internet access became an issue of political policy.

1990s

Universal Service Fund: The Universal Service Fund (USF) was created in 1997 by the United States FCC. It was established in order to meet the funding criteria and goals mandated in the Telecommunications Act of 1996. This act required all telecommunications companies to contribute money to the USF. The Universal Service Administrative Company (USAC), is responsible for allocating money from the central fund to four programs:

- High Cost
- Low Income
- Rural Health Care
- Schools & Libraries.

These four programs illustrate the goals of the Universal Service Fund, which are:

1. to promote the availability of quality services at just, reasonable, and affordable rates
2. to increase access to advanced telecommunications services throughout the Nation
3. to advance the availability of such services to all consumers, including those in low income, rural, insular, and high cost areas at rates that are reasonably comparable to those charged in urban areas.

Telecommunications companies may attempt to recover funds they pay into the Universal Service fund by charging their customers a fee. Consumers may see this practice reflected in a line-item charge on telecommunications bills labelled "Universal Service." The method and amount of this charge is a decided by the companies and is not required by the FCC.

E-Rate

Also known as the Schools and Libraries Program, E-Rate is funded by the Universal Service Fund. E-Rate was started in order to provide discounts to schools and libraries in the United States, so that they would be able to request funds to subsidize telecommunications services, Internet access, internal connections, and basic maintenance of internal connections. Depending on the poverty level and urban/rural status of the affected population, funds provided through E-Rate cover 20% to 90% of these costs. There is a fair amount of controversy surrounding E-Rate's use and management of funds. This controversy has included legal challenges from states and telecommunications companies. At present, the impact of the program is unclear as it is hard to measure.

Leading Broadband Companies in the United States

Even though there are numerous cable broadband companies in the United States, only three are seen as the main providers. Comcast, Time Warner and Cox Communications provide a majority of the cable broadband subscribers. Even without the cable market, Comcast is still seen as one of the leading Broadband providers. Still, the leading broadband provider in the United States is SBC Communications (AT&T) with 14.8 million subscribers as of the 2008 third quarter.

Implementing a National Broadband Policy

Availability of Internet access has rapidly become a vital tool in development and social progress since the turn of the century. As a result, broadband penetration rates are now treated as key economic indicators. The term "broadband penetration" refers to the percentage of the Internet access market that high speed has captured in a single country.

The United States has over 67.7 million people subscribed to the top broadband providers, which account for 94% of the market. In 2008, over 5.4 million high speed Internet subscribers were added, compared to 8.5 million in 2007. The market's peak subscriber additions were acquired in 2006 with 10.4 million; the slowed growth can be attributed to increased market penetration, according to the Leichtman Research Group (LRG) that has been tracking the broadband industry.. In March 2009, broadband penetration in active Internet user US homes dropped to 93.13%, creating the second consistent decrease from its peak of 93.38% in January 2009.

In 2008, the United States ranked 15th out of 30 countries in broadband penetration rates. This low worldwide ranking is surprising to many, as the Internet itself was invented in the United States. The country places behind most other developed nations, including the UK, Germany, France, Denmark, Switzerland, and Canada. The ranking also fuels a debate around the need for a national broadband policy, which would provide high speed broadband internet access to all citizens. The difficulty of successfully designating and distributing government funds in order to increase Internet access (particularly via broadband) are central limiting factors in the development of such a policy, but proponents believe that establishing one is necessary for social and economic progress. Those demanding a national broadband policy argue that such a policy is the best method by which the United States could achieve "universal availability and adoption of truly high-speed access.

National Broadband Policy-Possible Advantages

General Public Needs Met

In the 21st century, Internet access has been recognized as a basic public utility like water and electricity. Most agree it is needed in order to fully participate in society. For example, in the 2008 presidential elections, candidates delivered weekly addresses through videos on the Internet. These videos were often not available anywhere else. Proponents of national broadband policy argue that situations like the 2008 presidential elections illustrate that broadband internet access is a vital commodity for complete civic engagement.

Though the presidential election is a far more important subject than media entertainment, broadband can still provide daily entertainment where television cannot. Entertainment such as Webisodes, which are mini-episodes that are not usually aired on cable television, are an example of this extension of television. For example, Bravo (US TV Network) has given users the opportunity to view "Top Chef" on their personal computers. An entertainment advantage of these Webisode is the idea that the users are able to view the behind the scenes footage that occur after the contestants are kicked off the show. This kind of extension of entertainment marks a changing in the way we utilize internet access, and a hybridization of media.

Supporters of national broadband policy also argue that Internet access allows consumers to contribute to economic growth and innovation. Many services have moved business onto the Internet, a shift that is manifested through tools such as online banking, account referencing for utilities, and online library databases.

Those in favour of a national broadband policy also claim that many educational resources, including online classes and interactive learning simulations can only be found on the web. They believe that Internet access for all is necessary to create a truly American environment of equal opportunity.

Proponents of national broadband policy argue that online library databases, textbooks, and encyclopedias, have made the Internet a premier tool for education and lifelong learning. They maintain that the availability of broadband for all would provide a number of other highly valuable public benefits.

- More citizens could research hospitals and physicians prior to travelling so as to receive better medical care (a benefit particularly important for those who live in rural areas and travel long distances for medical treatment).
- More citizens could research jobs and investments, providing more opportunities for economic stability.
- More citizens could work from home, which accommodates personal or health needs, as well as reduces the use of travel-related resources usually used for commuting.
- More citizens could educate themselves on political issues, news, and other important civic issues, using

the Internet to become more informed and/or more active.

- If all citizens had Internet access, the government would have a more efficient tool to communicate and interact with its citizens quickly and effectively, particularly in terms of processes currently executed on paper. Broadband could also help the government provide aid efficiently, particularly in areas affected by disasters.

Rural Access

Some in favour of a national broadband policy argue it will provide the infrastructure necessary to bring broadband to rural areas of the United States. Many inhabitants of rural areas who are willing to pay for broadband service cannot get it. Low population density causes high costs and low profits and prevents most providers from doing business in the rural United States.

The American Recovery and Reinvestment Act of 2009, also known as the Stimulus Bill of 2009, provides $2.5 billion in the United States Department of Agriculture's Rural Utilities Service (RUS) for expanding rural broadband services. The RUS already helps to provide other utilities to rural areas of the U.S. through public-private partnerships. The money from the 2009 bill will be used to lay more fiber optic cables and hire more network specialists, engineers, and broadband company workers.

Urban Access

Proponents of national broadband policy believe that it would have a positive effect on the urban poor and provide them with many more opportunities for Internet access. In impoverished urban areas, many people are unable to afford the payments to bring broadband services into their homes. In 2008, roughly 50% of the United States population made an annual income of $50,000 or less. Of that half of the population, only 35% of the population could afford cost of monthly payments for broadband services. Some telecommunications companies have made a small effort to provide more affordable broadband rates for the 15% of the population that could not afford broadband before. However, non-profits like One Economy are the current primary source of funding used to provide low-income families with broadband access.

National Broadband Policy-Possible Disadvantages

Generally, those against a national broadband policy are not as outspoken as those in favour of one. The allocation of funds to broadband development included in the American Recovery and Reinvestment Act of 2009 was, however, met with voiced disapproval and questioning by opponents of such a national broadband policy.

Possible Negative Effect on Economy

Those against a national broadband policy argue that allowing the government to regulate and manage broadband development will cause efficiency and progress to wane, due to lack of competition. They also maintain that government involvement will inevitably result in profit losses for companies and businesses. Opponents of national broadband policy warn that in an attempt to achieve the desired goal of broadband for all, the government will waste billions of dollars on an infrastructure that people do not need or cannot use. They believe the money will be ill-spent because the government will not give attention to customer needs and desires with the same specificity exercised by the corporate sector.

Furthermore, those who do not advocate a national broadband policy claim that increases in broadband coverage could have an unexpected negative effect on the United States economy. They believe that many service sector jobs currently based in the United States could be moved offshore and fulfilled by lower-wage foreign workers. Opponents thus suggest that the relationship between economic growth and broadband penetration is questionable, emphasizing the notion that statistics are often misleading. They do not believe that such statistics are an adequate indicator that a national broadband policy is needed in the United States.

American Recovery and Reinvestment Act of 2009

The 2009 Stimulus Bill, as it is commonly termed, was enacted by the 111th United States Congress and signed into law by President Barack Obama on February 17, 2009. This bill provides money to create the Broadband Technology Opportunities Program through the National Telecommunications and Information Administration within the Department of Commerce.

- $4.7 billion to bring broadband to unserved and under served areas and to facilitate broadband use and adoption.
- $2.5 billion to the Federal Communications Commission to facilitate the development of a national broadband plan with one year.

The Federal Communications Commission solicited public comments in from April, 2009 through February, 2010. The plan was published in March, 2010.

Internet Service Providers (ISPs): 7,600 (1999 est.)— Because of aggressive lobbying and the United States' strong libertarian traditions, the Internet service provider industry remains relatively unregulated in comparison to other communications industries.

Country code (Top level domain): US— For various historical reasons, the.us domain was never widely used outside of a small number of government agencies and school districts. Most companies signed up for top level domains like.com instead.

NeuStar Inc. now has control over the.us registry and is trying to promote the domain as an option for American-oriented Web sites.

5

Good Neighbour in the North: Canada

Canada is a country occupying most of northern North America, extending from the Atlantic Ocean in the east to the Pacific Ocean in the west and northward into the Arctic Ocean. It is the world's second largest country by total area. Canada's common border with the United States to the south and northwest is the longest in the world.

The land occupied by Canada was inhabited for millennia by various groups of Aboriginal peoples. Beginning in the late 15th century, British and French expeditions explored, and later settled, along the Atlantic coast. France ceded nearly all of its colonies in North America in 1763 after the Seven Years' War. In 1867, with the union of three British North American colonies through Confederation, Canada was formed as a federal dominion of four provinces. This began an accretion of provinces and territories and a process of increasing autonomy from the United Kingdom. This widening autonomy was highlighted by the Statute of Westminster of 1931 and culminated in the Canada Act of 1982, which severed the vestiges of legal dependence on the British parliament.

A federation consisting of ten provinces and three territories, Canada is governed as a parliamentary democracy and a constitutional monarchy with Queen Elizabeth II as its head of state. It is a bilingual nation with both English and French as official languages at the federal level. One of the world's highly developed countries, Canada has a diversified economy that is reliant upon its abundant natural resources and upon trade—particularly with the United States, with which Canada has had a long and complex relationship. It is a member of the G8, G-20, NATO, OECD, WTO, Commonwealth, Francophonie, OAS, APEC, and UN.

Etymology

The name Canada comes from a St. Lawrence Iroquoian word, kanata, meaning "village" or "settlement". In 1535, indigenous inhabitants of the present-day Quebec City region used the word to direct French explorer Jacques Cartier towards the village of Stadacona. Cartier later used the word Canada to refer not only to that particular village, but also the entire area subject to Donnacona (the chief at Stadacona); by 1545, European books and maps had begun referring to this region as Canada. From the early 17th century onwards, that part of New France that lay along the Saint Lawrence River and the northern shores of the Great Lakes was known as Canada. The area was later split into two British colonies, Upper Canada and Lower Canada. They were re-unified as the Province of Canada in 1841. Upon Confederation in 1867, the name Canada was adopted as the legal name for the new country, and Dominion (a term from Psalm 72:8) was conferred as the country's title. Combined, the term Dominion of Canada was in common usage until the 1950s. As Canada asserted its political autonomy from the United Kingdom, the federal government increasingly used simply Canada on state documents and treaties, a change that was reflected in the renaming of the national holiday from Dominion Day to Canada Day in 1982.

Aboriginal Peoples

Aboriginal peoples in Canada include the First Nations, Inuit, and Metis. The descriptors "Indian" and "Eskimo" are falling into disuse. Archaeological and Indigenous genetic studies support a human presence in the northern Yukon from 26,500 years ago, and in southern Ontario from 9,500 years ago. Old Crow Flats and Bluefish Caves are two of the earliest archaeological sites of human (Paleo-Indians) habitation in Canada. Among the First Nations peoples, there are eight unique stories of creation and their adaptations.

These are the earth diver, world parent, emergence, conflict, robbery, rebirth of corpse, two creators and their contests, and the brother myth. The characteristics of Canadian Aboriginal civilizations included permanent or urban settlements, agriculture, civic and monumental architecture, and complex societal hierarchies. Some of these civilisations had long faded by the time of the first permanent European arrivals (c. late 15th–early 16th centuries), and have been discovered through archaeological investigations. The aboriginal population is estimated to have been between 200,000 and two million in the late 1400s. Repeated outbreaks of European infectious diseases such as influenza, measles and smallpox (to which they had no natural immunity), combined with other effects of European contact, resulted in an eighty-five to ninety-five percent aboriginal population decrease post-contact. The Metis culture of mixed blood originated in the mid-17th century when First Nation and Inuit married European settlers. The Inuit had more limited interaction with European settlers during the early periods.

European Colonization

Europeans first arrived when Norse sailors (often referred to as Vikings) settled briefly at L'Anse aux Meadows in Newfoundland around AD 1000; after the failure of that colony, there was no known further attempt at Canadian exploration until 1497, when Italian seafarer Giovanni Caboto (John Cabot) explored Canada's Atlantic coast for England. In 1534 Jacques Cartier explored Canada for France. French explorer Samuel de Champlain arrived in 1603 and established the first permanent European settlements at Port Royal in 1605 and Quebec City in 1608. Among French colonists of New France, Canadiens extensively settled the Saint Lawrence River valley and Acadians settled the present-day Maritimes, while French fur traders and Catholic missionaries explored the Great Lakes, Hudson Bay, and the Mississippi watershed to Louisiana. The French and Iroquois Wars broke out over control of the fur trade.

The English established fishing outposts in Newfoundland around 1610 and established the Thirteen Colonies to the south. A series of four Intercolonial Wars erupted between 1689 and

1763. Mainland Nova Scotia came under British rule with the Treaty of Utrecht (1713); the Treaty of Paris (1763) ceded Canada and most of New France to Britain after the Seven Years' War.

The Royal Proclamation (1763) carved the Province of Quebec out of New France and annexed Cape Breton Island to Nova Scotia. St. John's Island (now Prince Edward Island) became a separate colony in 1769. To avert conflict in Quebec, the British passed the Quebec Act of 1774, expanding Quebec's territory to the Great Lakes and Ohio Valley. It re-established the French language, Catholic faith, and French civil law there. This angered many residents of the Thirteen Colonies and helped to fuel the American Revolution.

The Treaty of Paris (1783) recognized American independence and ceded territories south of the Great Lakes to the United States. Around 50,000 United Empire Loyalists fled the United States to Canada. New Brunswick was split from Nova Scotia as part of a reorganization of Loyalist settlements in the Maritimes. To accommodate English-speaking Loyalists in Quebec, the Constitutional Act of 1791 divided the province into French-speaking Lower Canada (later the province of Quebec) and English-speaking Upper Canada (later Ontario), granting each its own elected Legislative Assembly.

Canada (Upper and Lower) was the main front in the War of 1812 between the United States and the British Empire. Following the war, large-scale immigration to Canada from Britain and Ireland began in 1815. From 1825 to 1846, 626,628 European immigrants landed at Canadian ports. Between one-quarter and one-third of all Europeans who immigrated to Canada before 1891 died of infectious diseases. The timber industry surpassed the fur trade in economic importance in the early nineteenth century.

The desire for responsible government resulted in the aborted Rebellions of 1837. The Durham Report subsequently recommended responsible government and the assimilation of French Canadians into British culture. The Act of Union 1840 merged The Canadas into a united Province of Canada. Responsible government was established for all British North

American provinces by 1849. The signing of the Oregon Treaty by Britain and the United States in 1846 ended the Oregon boundary dispute, extending the border westward along the 49th parallel. This paved the way for British colonies on Vancouver Island (1849) and in British Columbia (1858). Canada launched a series of exploratory expeditions to claim Rupert's Land and the Arctic region.

Confederation and Expansion

Following several constitutional conferences, the Constitution Act, 1867 officially proclaimed Canadian Confederation, creating "one Dominion under the name of Canada" on July 1, 1867, with four provinces: Ontario, Quebec, Nova Scotia, and New Brunswick. Canada assumed control of Rupert's Land and the North-Western Territory to form the Northwest Territories, where the Metis' grievances ignited the Red River Rebellion and the creation of the province of Manitoba in July 1870. British Columbia and Vancouver Island (which had united in 1866) and the colony of Prince Edward Island joined the Confederation in 1871 and 1873, respectively. Prime Minister John A. Macdonald's Conservative government established a national policy of tariffs to protect nascent Canadian manufacturing industries.

To open the West, the government sponsored construction of three trans-continental railways (including the Canadian Pacific Railway), opened the prairies to settlement with the Dominion Lands Act, and established the North-West Mounted Police to assert its authority over this territory. In 1898, after the Klondike Gold Rush in the Northwest Territories, the Canadian government created the Yukon territory. Under Liberal Prime Minister Wilfrid Laurier, continental European immigrants settled the prairies, and Alberta and Saskatchewan became provinces in 1905.

Early 20th Century

Because Britain still maintained control of Canada's foreign affairs under the Confederation Act, its declaration of war in 1914 automatically brought Canada into World War I. Volunteers sent to the Western Front later became part of the Canadian Corps. The Corps played a substantial role in the Battle of Vimy Ridge and other major battles of the war. Out of approximately 625,000 who served, about 60,000 were killed and another 173,000 were wounded. The Conscription Crisis of 1917 erupted when conservative Prime Minister Robert Borden brought in compulsory military service over the objection of French-speaking Quebecers. In 1919, Canada joined the League of Nations independently of Britain and in 1931, the Statute of Westminster affirmed Canada's independence.

The Great Depression brought economic hardship all over Canada. In response, the Co-operative Commonwealth Federation (CCF) in Alberta and Saskatchewan enacted many measures of a welfare state (as pioneered by Tommy Douglas) into the 1940s and 1950s. Canada declared war on Germany independently during World War II under Liberal Prime Minister William Lyon Mackenzie King, three days after Britain. The first Canadian Army units arrived in Britain in December 1939.

Canadian troops played important roles in the failed 1942 Dieppe Raid in France, the Allied invasion of Italy, the D-Day landings, the Battle of Normandy, and the Battle of the Scheldt in 1944. Canada provided asylum and protection for the monarchy of the Netherlands while that country was occupied, and is credited by the country for leadership and major contribution to its liberation from Nazi Germany. The Canadian economy boomed as industry manufactured military materiel for Canada, Britain, China, and the Soviet Union. Despite another Conscription Crisis in Quebec, Canada finished the war with one of the largest armed forces in the world, and the second-wealthiest economy. In 1945, during the war, Canada became one of the founding members of the United Nations.

Modern Times

The Dominion of Newfoundland (now Newfoundland and Labrador), at the time equivalent in status to Canada and Australia as a Dominion, joined Canada in 1949. Canada's growth, combined with the policies of successive Liberal governments, led to the emergence of a new Canadian identity, marked by the adoption of the current Maple Leaf Flag in 1965, the implementation of official bilingualism (English and French) in 1969, and official multiculturalism in 1971. There was also the founding of socially democratic programmes, such as universal health care, the Canada Pension Plan, and Canada Student Loans, though provincial governments, particularly Quebec and Alberta, opposed many of these as incursions into their jurisdictions. Finally, another series of constitutional conferences resulted in the 1982 patriation of Canada's constitution from the United Kingdom, concurrent with the creation of the Charter of Rights and Freedoms. In 1999, Nunavut became Canada's third territory after a series of negotiations with the federal government.

At the same time, Quebec was undergoing profound social and economic changes through the Quiet Revolution, giving birth to a nationalist movement in the province and the more radical Front de liberation du Quebec (FLQ), whose actions ignited the October Crisis in 1970. A decade later, an unsuccessful referendum on sovereignty-association was held in 1980, after which attempts at constitutional amendment failed in 1990. A second referendum followed in 1995, in which sovereignty was rejected by a slimmer margin of just 50.6% to 49.4%. In 1997, the Supreme Court ruled that unilateral secession by a province would be unconstitutional, and the Clarity Act was passed by parliament, outlining the terms of a negotiated departure from Confederation.

In addition to the issues of Quebec sovereignty, a number of crises shook Canadian society in the late 1980s and early 1990s. These included the explosion of Air India Flight 182 in 1985, the largest mass murder in Canadian history; the Ecole Polytechnique massacre in 1989, a university shooting targeting female students; and the Oka Crisis in 1990, the first of a number of violent confrontations between the government and Aboriginal groups. Canada also joined the Gulf War in 1990 as part of a US-led coalition force, and was active in several peacekeeping missions in the late 1990s. It sent troops to Afghanistan in 2001, but declined to send forces to Iraq when the US invaded in 2003.

Government and Politics

Canada has strong democratic traditions upehld through a parliamentary government within the construct of constitutional monarchy, the monarchy of Canada being the foundation of the executive, legislative, and judicial branches and its authority stemming from the Canadian populace. The sovereign is Queen Elizabeth II, who also serves as head of state of 15 other Commonwealth countries and resides predominantly in the United Kingdom. As such, the Queen's representative, the Governor General of Canada (presently Michaelle Jean), carries out most of the royal duties in Canada.

The direct participation of the royal and viceroyal figures in any of these areas of governance is limited, though; in practice, their use of the executive powers is directed by the Cabinet, a committee of ministers of the Crown responsible to the elected House of Commons and headed by the Prime Minister of Canada (presently Stephen Harper), the head of government. To ensure the stability of government, the governor general will usually appoint as prime minister the person who is the current leader of the political party that can obtain the confidence of a plurality in the House of Commons and the prime minister chooses the Cabinet. The Prime Minister's Office (PMO) is thus one of the most powerful institutions in government, initiating most legislation for parliamentary approval and selecting for appointment by the Crown, besides the aforementioned, the governor general, lieutenant governors, senators, federal court judges, and heads of crown corporations and government agencies. The leader of the party with the second-most seats usually becomes the Leader of Her Majesty's Loyal Opposition (presently Michael Ignatieff) and is part of an adversarial parliamentary system intended to keep the government in check.

Each Member of Parliament in the House of Commons is elected by simple majority in an electoral district or riding. General elections must be called by the governor general, on the advice of the prime minister, within four years of the previous election, or may be triggered by the government losing a confidence vote in the House. Members of the Senate, whose seats are apportioned on a regional basis, serve until age 75. Four parties had representatives elected to the federal parliament in the 2008 elections: the Conservative Party of Canada (governing party), the Liberal Party of Canada (the Official Opposition), the New Democratic Party (NDP), and the Bloc Quebecois. The list of historical parties with elected representation is substantial.

Canada's federal structure divides government responsibilities between the federal government and the ten provinces. Provincial legislatures are unicameral and operate in parliamentary fashion similar to the House of Commons. Canada's three territories also have legislatures, but these are not sovereign and have fewer constitutional responsibilities than the provinces and with some structural differences.

Law

The Constitution of Canada is the supreme law of the country, and consists of written text and unwritten conventions. The Constitution Act, 1867 (known as the British North America Act prior to 1982) affirmed governance based on parliamentary precedent "similar in principle to that of the United Kingdom" and divided powers between the federal and provincial governments; the Statute of Westminster, 1931 granted full autonomy; and the Constitution Act, 1982 added the Canadian Charter of Rights and Freedoms, which guarantees basic rights and freedoms that usually cannot be overridden by any level of government—though a notwithstanding clause allows the federal parliament and provincial legislatures to override certain sections of the Charter for a period of five years—and added a constitutional amending formula.

Although not without conflict, European Canadians' early interactions with First Nations and Inuit populations were relatively peaceful. Combined with Canada's late economic development in many regions, this peaceful history has allowed Canadian Indigenous peoples to have a relatively strong influence on the national culture while preserving their own identity. The Canadian Crown and Aboriginal peoples began interactions during the European colonialisation period. Numbered treaties, the Indian Act, the Constitution Act of 1982 and case laws were established. A series of eleven treaties were signed between Aboriginals in Canada and the reigning Monarch of Canada from 1871 to 1921. These treaties are agreements with the Government of Canada administered by Canadian Aboriginal law and overseen by the Minister of Indian Affairs and Northern Development. The role of the treaties was reaffirmed by Section Thirty-five of the Constitution Act, 1982, which "recognizes and affirms existing Aboriginal and treaty rights". These rights may include provision of services such as health care, and exemption from taxation. The legal and policy framework within which Canada and First Nations operate was further formalised in 2005, through the First Nations–Federal Crown Political Accord, which established cooperation as "a cornerstone for partnership between Canada and First Nations".

Canada's judiciary plays an important role in interpreting laws and has the power to strike down laws that violate the Constitution. The Supreme Court of Canada is the highest court and final arbiter and has been led by the Right Honourable Madam Chief Justice Beverley McLachlin, P.C. (the first female Chief Justice) since 2000. Its nine members are appointed by the governor general on the advice of the Prime Minister and Minister of Justice. All judges at the superior and appellate levels are appointed after consultation with nongovernmental legal bodies. The federal cabinet also appoints justices to superior courts at the provincial and territorial levels. Judicial posts at the lower provincial and territorial levels are filled by their respective governments.

Common law prevails everywhere except in Quebec, where civil law predominates. Criminal law is solely a federal responsibility and is uniform throughout Canada. Law enforcement, including criminal courts, is a provincial responsibility, but in rural areas of all provinces except Ontario and Quebec, policing is contracted to the federal Royal Canadian Mounted Police.

Foreign Relations and Military

Canada and the United States share the world's longest undefended border, co-operate on military campaigns and exercises, and are each other's largest trading partner. Canada nevertheless has an independent foreign policy, most notably maintaining full relations with Cuba and declining to participate in the Iraq War. Canada also maintains historic ties to the United Kingdom and France and to other former British and French colonies through Canada's membership in the Commonwealth of Nations and the Francophonie. Canada is noted for having a strong and positive relationship with the Netherlands, and the Dutch government traditionally gives tulips, a symbol of the Netherlands, to Canada each year in remen rance of the latter country's contribution to its liberation.

Canada currently employs a professional, volunteer military force of about 67,000 regular and 26,000 reserve personnel. The unified Canadian Forces (CF) comprise the army, navy, and air force.

Canada is an industrial nation with a highly developed science and technology sector. Since the First World War, Canada has produced its own infantry fighting vehicle, anti-tank guided missile and small arms for the Canadian Forces and particularly for the army. The Canadian Forces operate state of the art equipments able to handle modern threats through 2030-2035. Despite the financial cut between 1987–2004, the Canadian Forces are well equipped. The Land Force Command currently operate approximatively 10 500 utility vehicles including G-wagon and 7000-MV and also operate approximatively 2 700 armoured fighting vehicles including the LAV-III and the Leopard 2. The land force also operate approximatively 150 field artillery including the M777 howitzer and the LG1 Mark II. The Canadian navy currently operates 33 combat vessels. These include the Halifax class frigate and the Victoria class submarine. The Canadian air force operates 333 aircraft. These include the CF-188 Hornet, CC-130 Hercule and the CH-146 Griffon.

Strong attachment to the British Empire and Commonwealth led to major participation in British military efforts in the Second Boer War, the First World War, and the Second World War. Since then, Canada has been an advocate for multilateralism, making efforts to resolve global issues in collaboration with other nations. Canada was a founding member of the United Nations in 1945 and of NATO in 1949. During the Cold War, Canada was a major contributor to UN forces in the Korean War and founded the North American Aerospace Defence Command (NORAD) in cooperation with the United States to defend against potential aerial attacks from the Soviet Union.

During the Suez Crisis of 1956, future Prime Minister Lester B. Pearson eased tensions by proposing the inception of the United Nations Peacekeeping Force, for which he was awarded the 1957 Nobel Peace Prize. As this was the first UN peacekeeping mission, Pearson is often credited as the inventor of the concept. Canada has since served in 50 peacekeeping missions, including every UN peacekeeping effort until 1989, and has since maintained forces in international missions in Rwanda, the former Yugoslavia, and elsewhere; Canada has sometimes faced controversy over its involvement in foreign countries, notably in the 1993 Somalia Affair. The number of Canadian military personnel participating in peacekeeping missions has decreased greatly in the past two decades. As of June 30, 2006, 133 Canadians were currently serving on United Nations peacekeeping missions worldwide, including 55 Canadian military personnel, compared with 1044 military personnel as of December 31, 1996.

Canada joined the Organization of American States (OAS) in 1990 and hosted the OAS General Assembly in Windsor, Ontario, in June 2000 and the third Summit of the Americas in Quebec City in April 2001. Canada seeks to expand its ties to Pacific Rim economies through membership in the Asia-Pacific Economic Cooperation forum (APEC).

Since 2001, Canada has had troops deployed in Afghanistan as part of the U.S. stabilization force and the UN-authorized, NATO-commanded International Security Assistance Force. Canada has committed to withdraw from Kandahar Province by 2011, by which time it will have spent an estimated total of $11.3 billion on the mission. Canada and the U.S. continue to integrate state and provincial agencies to strengthen security along the Canada-United States border through the Western Hemisphere Travel Initiative.

In February 2007, Canada, Italy, Britain, Norway, and Russia announced their funding commitments to launch a $1.5 billion project to help develop vaccines they said could save millions of lives in poor nations, and called on others to join them. In August 2007, Canadian sovereignty in Arctic waters was challenged after a Russian underwater expedition to the North Pole; Canada has considered that area to be sovereign territory since 1925.

Provinces and Territories

The provinces and territories of Canada combine to make

up the world's second largest country by area. There are ten provinces and three territories in all. The major difference between a Canadian province and a territory is that provinces are jurisdictions that receive their power and authority directly from the Constitution Act, 1867, whereas territories derive their mandates and powers from the federal government.

Currently, the ten provinces are Alberta, British Columbia, Manitoba, New Brunswick, Newfoundland and Labrador, Nova Scotia, Ontario, Prince Edward Island, Quebec, and Saskatchewan, while the three territories are Northwest Territories, Nunavut, and Yukon.

Location of Provinces and Territories

History

Ontario, Quebec, New Brunswick, and Nova Scotia are the original provinces, formed when British North American colonies federated on July 1, 1867, into the Dominion of Canada and by stages began accruing the indicia of sovereignty from the United Kingdom. Ontario and Quebec were united before Confederation as the Province of Canada. Over the following six years, Manitoba, British Columbia, and Prince Edward Island were added as provinces.

The Hudson's Bay Company maintained control of large swathes of western Canada until 1870, when it turned over the land to the Government of Canada, forming part of the Northwest Territories. Manitoba and the Northwest Territories were created in 1870 from Rupert's Land and the North-Western Territory. At the time, the land comprising the Northwest Territories was all of current northern and western Canada, including the northern two thirds of Ontario and Quebec, with exception of the Arctic Islands, British Columbia and a small portion of southern Manitoba. On September 1, 1905, a portion of the Northwest Territories south of the 60° parallel became the provinces of Alberta and Saskatchewan. In 1912, the boundaries of Quebec, Ontario and Manitoba were expanded northward: Manitoba's to the 60° parallel, Ontario's to Hudson Bay and Quebec's to encompass the District of Ungava.

In 1869, the people of Newfoundland voted to remain a British colony over concerns that central Canada would dominate taxation and economic policy. In 1907, Newfoundland acquired dominion status. In 1933, facing national bankruptcy, the legislature turned over political control to the Commission of Government. Following World War II, in a 1948 referendum, a narrow majority of Newfoundland citizens voted to join Confederation and, on March 31, 1949, Newfoundland became Canada's tenth and final province. In 2001 it was officially renamed Newfoundland and Labrador.

In 1903, the Alaska Panhandle Dispute fixed British Columbia's northwestern boundary. This was one of only two provinces in Canadian history to have its size reduced. The second, in 1927, occurred when a boundary dispute between the province of Quebec and the Dominion of Newfoundland saw Labrador increased at Quebec's expense.

In 1999, Nunavut was created from the eastern portion of the Northwest Territories. Yukon lies in the western portion of The North, while Nunavut is in the east.

All three territories combined are the most sparsely populated region in Canada with about 100,000 people spread across a huge area. They are often referred to as a single region, The North, for organisational purposes. The District of Keewatin was created as a separate territory from 1876 to 1905, after which, as the Keewatin Region, it became an administration district of the Northwest Territories. In 1999, it was dissolved when it became part of Nunavut.

In late 2004, Prime Minister Paul Martin surprised some observers by expressing his personal support for all three territories gaining provincial status "eventually". He cited their importance to the country as a whole and the ongoing need to assert sovereignty in the Arctic, particularly as global warming could make that region more open to exploitation.

Government

Theoretically, provinces have a great deal of power relative to the federal government, with jurisdiction over many public goods such as healthcare, education, welfare, and intra-provincial transportation. They receive "transfer payments" from the federal government to pay for these, as well as exacting their own taxes. In practice, however, the federal government can use these transfer payments to influence these provincial areas. For instance in order to receive healthcare funding under medicare, provinces must agree to meet certain federal mandates, such as universal access to required medical treatment.

Provincial and territorial legislatures have no second chamber like the Canadian Senate. Originally, most provinces did have such bodies, known as legislative councils, but these were subsequently abolished, Quebec's being the last in 1968. In most provinces, the single house of the legislature is known as the Legislative Assembly except in Nova Scotia and Newfoundland and Labrador, where it is called the House of Assembly, and Quebec where it is generally called the National Assembly. Ontario has a Legislative Assembly but its members are called Members of the Provincial Parliament or MPPs. The legislative assemblies use a procedure similar to that of the Canadian House of Commons. The head of government of each province, called the premier, is generally the head of the party with the most seats. This is also the case in Yukon, but the Northwest Territories and Nunavut have no political parties at the territorial level. The Queen's representative to each province is the Lieutenant Governor. In each of the territories there is an analogous Commissioner, but he or she represents the federal government and not the monarch per se.

Each of the territories elects one Member of Parliament. Canadian territories are each entitled to elect one full voting representative to the Canadian House of Commons. With the sole exception of Prince Edward Island having slightly greater per capita representation than the Northwest Territories, every territory has considerably greater per capita representation in the Commons than every other province. Residents of the Canadian territories are full citizens and enjoy the same rights as all other Canadians. Each territory also has one Senator.

Provincial Parties

Most provinces have provincial counterparts to the three national federal parties. However, some provincial parties are not formally linked to the federal parties that share the same name. The New Democratic Party is the only party that has integrated membership between the provincial and federal wings. Some provinces have regional political parties, such as the Saskatchewan Party. The provincial political climate of Quebec is quite different: the main split is between sovereignty, represented by the Parti Quebecois, and federalism, represented primarily by the Quebec Liberal Party. From March 2007 to December 2008, the Official Opposition was the Action democratique du Quebec, which advocates what it calls "autonomy", a middle-of-the-road option supporting localised power in the Federal structure. They have no corresponding Federal party, but polls show their base to align with the Federal Conservative Party of Canada.

The provincial Progressive Conservative parties are also now separate from the federal Conservative Party, which resulted from a merger between the Progressive Conservatives and the Canadian Alliance. In British Columbia, the Liberal Party separated from the federal Liberal Party and is now an independent entity.

Canada is a federation composed of ten provinces and three territories. In turn, these may be grouped into regions: Western Canada, Central Canada, Atlantic Canada, and Northern Canada (the latter made up of the three territories: Yukon, Northwest Territories, and Nunavut). Eastern Canada refers to Central Canada and Atlantic Canada together. Provinces have more autonomy than territories. The provinces are responsible for most of Canada's social programs (such as health care, education, and welfare) and together collect more revenue than the federal government, an almost unique structure among federations in the world. Using its spending powers, the federal government can initiate national policies in provincial areas, such as the Canada Health Act; the provinces can opt out of these, but rarely do so in practice. Equalization payments are made by the federal government to ensure that reasonably uniform standards of services and taxation are kept between the richer and poorer provinces.

Geography and Climate

Canada occupies a major northern portion of North America, sharing the land borders with the contiguous United States to the south and the U.S. state of Alaska to the northwest, stretching from the Atlantic Ocean in the east to the Pacific Ocean in the west; to the north lies the Arctic Ocean. By total area (including its waters), Canada is the second-largest country in the world—after Russia. By land area, Canada ranks fourth (land area is total area minus the area of lakes and rivers).

Since 1925, Canada has claimed the portion of the Arctic between 60°W and 141°W longitude, but this claim is not universally recognized. The northernmost settlement in Canada (and in the world) is Canadian Forces Station Alert on the northern tip of Ellesmere Island—latitude 82.5°N—817 kilometres (450 nautical miles, 508 miles) from the North Pole. Much of the Canadian Arctic is covered by ice and permafrost. Canada also has the longest coastline in the world: 202,080 kilometres (125,570 mi).

The population density, 3.3 inhabitants per square kilometre (8.5/sq mi), is among the lowest in the world. The most densely populated part of the country is the Quebec City – Windsor Corridor, (situated in Southern Quebec and Southern Ontario) along the Great Lakes and the Saint Lawrence River in the southeast.

Canada has an extensive coastline on its north, east, and west, and since the last glacial period it has consisted of eight distinct forest regions, including extensive boreal forest on the Canadian Shield. The vastness and variety of Canada's geography, ecology, vegetation and land forms have given rise to a wide variety of climates throughout the country. Because of its vast size, Canada has more lakes than any other country, containing much of the world's fresh water. There are also fresh-water glaciers in the Canadian Rockies and the Coast Mountains.

Average winter and summer high temperatures across Canada vary according to the location. Winters can be harsh in many regions of the country, particularly in the interior and Prairie provinces, which experience a continental climate, where daily average temperatures are near "15 °C (5 °F) but can drop below "40 °C ("40.0 °F) with severe wind chills. In noncoastal regions, snow can cover the ground almost six months of the year (more in the north). Coastal British Columbia enjoys a temperate climate, with a mild and rainy winter. On the east and west coasts, average high temperatures are generally in the low 20s °C (70s °F), while between the coasts, the average summer high temperature ranges from 25 to 30 °C (77 to 86 °F), with occasional extreme heat in some interior locations exceeding 40 °C (104 °F).

Canada is also geologically active, having many earthquakes and potentially active volcanoes, notably Mount Meager, Mount Garibaldi, Mount Cayley, and the Mount Edziza volcanic complex. The volcanic eruption of Tseax Cone in 1775 caused a catastrophic disaster, killing 2,000 Nisga'a people and destroying their village in the Nass River valley of northern British Columbia; the eruption produced a 22.5-kilometre (14.0 mi) lava flow, and according to legend of the Nisga'a people, it blocked the flow of the Nass River.

Science and Technology

Canada is an industrial nation with a highly developed science and technology sector. Nearly 1.88% of Canada's GDP is allocated to research & development (R&D). The country has eighteen Nobel laureates in physics, chemistry and medicine. Canada ranks 12 in the world for Internet usage with 28.0 million users, 84.3% of the total population.

The Defence Research and Development Canada is an agency of the Department of National Defence,whose purpose is to respond to the scientific and technological needs of the Canadian Forces. Over the year DRDC have been responsible for numerous innovations and inventions of practical application both in civilian and military world. These include the CADPAT, G-suit, CRV7, Carbon dioxide laser and the Flight data recorder. DRDC also contribute in the development of the most advanced Active Electronically Scanned Array in the world as part of an international effort involving Canada, Germany, and the Netherlands.

The Canadian Space Agency conducts space, planetary, and aviation research, as well as develops rockets and satellites. In 1984, Marc Garneau became Canada's first astronaut, serving as payload specialist of STS-41-G. Canada is a participant in the International Space Station and one of the world's pioneers in space robotics with the Canadarm, Canadarm and Dextre. Canada was ranked third among 20 top countries in space sciences. Since the 1960s, Canada Aerospace Industries have designed and built 10 satellites, including RADARSAT-1, RADARSAT-2 and MOST. Canada also produced one of the most successful sounding rockets, the Black Brant; over 1000 have been launched since they were initially produced in 1961. Universities across Canada are working on the first domestic landing spacecraft: the Northern Light, designed to search for life on Mars and investigate Martian electromagnetic radiation environment and atmospheric properties. If the Northern Light is successful, Canada will be the third country to land on another planet.

Economy

Canada has the tenth largest economy in the world (measured in US dollars at market exchange rates), is one of the world's wealthiest nations, and is a member of the Organization for Economic Co-operation and Development (OECD) and Group of Eight (G8). As with other developed nations, the Canadian economy is dominated by the service industry, which employs about three quarters of Canadians. Canada is unusual among developed countries in the importance of the primary sector, with the logging and oil industries being two of Canada's most important. Canada also has a sizable manufacturing sector, centered in Central Canada, with the automobile industry especially important.

Canada has one of the highest levels of economic freedom in the world. Today Canada closely resembles the U.S. in its market-oriented economic system, and pattern of production. As of June 2010, Canada's national unemployment rate stood at 8.1% (or 7.2% when measured according to equivalent US concepts) as the economy continues it's recovery from the effects of the 2007-2010 global financial crisis. In May 2010, provincial unemployment rates varied from a low of 5.0% in Saskatchewan to a high of 13.8% in Newfoundland and Labrador. According to the Forbes Global 2000 list of the world's largest companies in 2008, Canada had 69 companies in the list, ranking 5th next to France. As of 2008, Canada's total government debt burden is the lowest in the G8.

International trade makes up a large part of the Canadian economy, particularly of its natural resources. In 2009, agricultural, energy, forestry and mining exports accounted for about 58% of Canada's total exports. Machinery, equipment, automotive products and other manufactures accounted for a further 38% of exports in 2009. In 2009, exports accounted for approximately 30% of Canada's GDP. The United States is by far its largest trading partner, accounting for about 73% of exports and 63% of imports as of 2009. Canada's combined exports and imports ranked 8th among all nations in 2006.

Canada has considerable natural resources spread across its varied regions. As an example, in British Columbia the forestry industry is of great importance, while the oil and gas industry is important in Alberta, Saskatchewan and Newfoundland and Labrador. Northern Ontario is home to a wide array of mines, while the fishing industry has long been central to the character of the Atlantic provinces, though it has recently been in steep decline. Canada has mineral resources of coal, copper, iron ore, and gold.

These primary industries are increasingly becoming relatively less important to the overall economy. Only some 4% of Canadians are employed in these fields, and they account for 6.2% of GDP. They are still paramount in many parts of the country. Many, if not most, towns in northern Canada, where agriculture is difficult, exist because of a nearby mine or source of timber. Canada is a world leader in the production of many natural resources such as gold, nickel, uranium, diamonds and lead. Several of Canada's largest companies are based in natural resource industries, such as EnCana, Cameco, Goldcorp, and Barrick Gold. The vast majority of these products are exported, mainly to the United States. There are also many secondary and service industries that are directly linked to primary ones. For instance one of Canada's largest manufacturing industries is the pulp and paper sector, which is directly linked to the logging industry.

The relatively large reliance on natural resources has several effects on the Canadian economy and Canadian society. While manufacturing and service industries are easy to standardize, natural resources vary greatly by region. This ensures that differing economic structures developed in each region of Canada, contributing to Canada's strong regionalism. At the same time the vast majority of these resources are exported, integrating Canada closely into the international economy. Howlett and Ramesh argue that the inherent instability of such industries also contributes to greater government intervention in the economy, to reduce the social impact of market changes.

Such industries also raise important questions of sustainability. Despite many decades as a leading producer, there is little risk of depletion. Large discoveries continue to be made, such as the massive nickel find at Voisey's Bay. Moreover the far north remains largely undeveloped as producers await higher prices or new technologies as many operations in this region are not yet cost effective. In recent decades Canadians have become less willing to accept the environmental destruction associated with exploiting natural resources. High wages and Aboriginal land claims have also curbed expansion. Instead many Canadian companies have focused their exploration and expansion activities overseas where prices are lower and governments more accommodating. Canadian companies are increasingly playing important roles in Latin America, Southeast Asia, and Africa.

The exploitation of renewable resources have raised concerns in recent years. After decades of escalating overexploitation the cod fishery all but collapsed in the 1990s, and the Pacific salmon industry also suffered greatly. The logging industry, after many years of activism, has in recent years moved to a more sustainable model.

Overview

Energy

Canada is one of the few developed nations that is a net exporter of energy-in 2009 net exports of energy products amounted to 2.9% of GDP. Most important are the large oil and gas resources centred in Alberta and the Northern Territories, but also present in neighbouring British Columbia and Saskatchewan. The vast Athabasca Oil Sands give Canada the world's second largest reserves of oil after Saudi Arabia according to USGS. In British Columbia and Quebec, as well as Ontario, Saskatchewan, Manitoba and the Labrador region, hydroelectric power is an inexpensive and relatively environmentally friendly source of abundant energy. In part because of this, Canada is also one of the world's highest per capita consumers of energy. Cheap energy has enabled the creation of several important industries, such as the large aluminum industry in Quebec, Alberta and British Columbia.

Historically, an important issue in Canadian politics is that while Western Canada is one of the world's richest sources of energy, the industrial heartland of Southern Ontario has fewer native sources of power. It is, however, cheaper for Alberta to ship its oil to the western United States than to eastern Canada. The eastern Canadian ports thus import significant quantities of oil from overseas, and Ontario makes significant use of nuclear power.

In times of high oil prices this means that the majority of Canada's population suffers, while the West benefits. The National Energy Policy of the early 1980s attempted to force Alberta to sell low priced oil to eastern Canada. This policy proved deeply divisive, and quickly lost its importance as oil prices collapsed in the mid-1980s. One of the most controversial sections of the Canada-United States Free Trade Agreement of 1988 was a promise that Canada would never charge the United States more for energy than fellow Canadians.

Petroleum Production in Canada

Petroleum production in Canada is a major industry which is important to the economy of North America. Canada is the seventh largest oil producing country in the world. In 2008 it produced an average of 438,000 cubic metres per day (2,750,000 bbl/d) of crude oil, crude bitumen and natural gas condensate. Of that amount, 45% was conventional crude oil, 49.5% was bitumen from oil sands, and 5.5% was condensate from natural gas wells. Most of Canadian petroleum production, approximately 283,000 cubic metres per day (1,780,000 bbl/d), was exported, almost all of it to the United States. Canada is the largest single source of oil imports into the United States.

The petroleum industry in Canada is also referred to as the Canadian "Oil Patch"; the term refers especially to upstream operations (exploration and production of oil and gas), and to a lesser degree to downstream operations (refining, distribution, and selling of oil and gas products). In 2005, almost 25,000 new oil wells were spud (drilled) in Canada. Daily, over 100 new wells are spud in the province of Alberta alone.

History

The Canadian petroleum industry developed in parallel with that of the United States. The first oil well in Canada was dug by hand (rather than drilled) in 1858 by James Miller Williams near his asphalt plant at Oil Springs, Ontario. At a depth of 20 metres (66 ft) he struck oil, one year before "Colonel" Edwin Drake drilled the first oil well in the United States. Williams later went on to found the "The Canadian Oil Company" which qualified as the world's first integrated oil company.

Petroleum production in Ontario expanded rapidly, and practically every significant producer became his own refiner. By 1864, 20 refineries were operating in Oil Springs and seven in Petrolia, Ontario. However, Ontario's status as an important oil producer did not last long. By 1880 Canada was a net importer of oil from the United States.

Canada's unique geography, geology, resources and patterns of settlement have been key factors in the history of Canada. The development of the petroleum sector helps illustrate how they have helped make the nation quite distinct from her neighbour to the south.

Divisions

Most exploration and production occurs in Alberta, with a significant number of operations in British Columbia—particularly in winter—and consistent activity in Saskatchewan. Drilling from large offshore platforms occurs on the Newfoundland continental shelf.

Alberta

Alberta is the largest producer of conventional crude oil, synthetic crude, natural gas and gas products in Canada. Two of the largest producers of petrochemicals in North America are located in central and north central Alberta. In both Red Deer and Edmonton, world class polyethylene and vinyl manufacturers produce products shipped all over the world, and Edmonton's oil refineries provide the raw materials for a large petrochemical industry to the east of Edmonton. There are hundreds of small companies in Alberta dedicated to providing all sorts of services to this industry—from drilling to well maintenance, pipeline maintenance to seismic exploration.

The Athabasca Oil Sands (previously known as the Athabasca Tar Sands) have estimated oil reserves in excess of that of the rest of the world, estimated to be 1.6 trillion barrels (254 km^3). With the advancement of extraction methods, bitumen and economical synthetic crude are produced at costs nearing that of conventional crude. This technology grew and developed in Alberta. Many companies employ both conventional strip mining and non-conventional methods to extract the bitumen from the Athabasca deposit. With current technology, only 315 billion barrels (50 km^3) are recoverable. Entire towns, like Fort McMurray, have grown up entirely because of the large multinational corporations which have taken on the task of oil production.

While Edmonton is considered the pipeline junction, manufacturing, chemical processing, research and refining centre of the province, Calgary is known for its senior and junior oil company head offices.

Major oil fields are found in southeast Alberta (Brooks, Medicine Hat, Lethbridge), northwest (Grande Prairie, High Level, Rainbow Lake, Zama), central (Caroline, Red Deer), and northeast (Athabasca Oil Sands).

Structural regions include: Foothills, Greater Arch, Deep Basin.

Oil and gas activity is regulated by the Alberta Energy Resources Conservation Board (ERCB) (Formerly the Energy and Utility Board (EUB)).

British Columbia

Drilling for gas and oil in the Peace Country of north-eastern British Columbia, around Fort Nelson (Greater Sierra oil field), Fort St. John (Pink Mountain, Border Ring) and Dawson Creek;

Oil and gas activity is regulated by the Oil and Gas Commission (OGC).

Manitoba: A few rigs drilling for gas in southern Manitoba

Saskatchewan: Mostly shallow gas wells in southwestern Saskatchewan (Hatton, Cypress Hill) and the southeast

(Lougheed, Weir Hill), heavy oil extraction around Lloydminster, oil wells around Weyburn.

Oil and gas activity is regulated by the Saskatchewan Industry and Resources (SIR).

Northern Canada (onshore): Drilling for oil in the Mackenzie Delta by Petro-Canada

Northern Canada (offshore): Production in the Beaufort Sea off the Mackenzie Delta.

Sporadic drilling along the continental shelf of the Beaufort Sea.

Eastern Canada (onshore): Sporadic drilling in southern Ontario by Talisman Energy Inc.

Sporadic drilling in western Newfoundland.

Sporadic drilling in northern Nova Scotia and western Cape Breton Island.

Sporadic drilling in northern and eastern Prince Edward Island.

Production of natural gas at the McCully Field in southern New Brunswick, connected to the Maritimes and Northeast Pipeline.

Eastern Canada (offshore): Offshore oil drilling and production at Hibernia, Terra Nova, and White Rose fields off the coast of Newfoundland

- Offshore gas drilling and production on Sable Island fields off the coast of Nova Scotia
- Sporadic drilling along continental shelf off Nova Scotia.
- Sporadic drilling in Laurentian Fan at southern end of Cabot Strait
- Sporadic drilling in eastern Northumberland Strait.

Major Players

The country's four largest integrated refiners are Imperial Oil, Husky Energy, Petro-Canada, and Suncor Energy. In 2007 Canada's four biggest oil companies brought in record profits of $11.75 billion, up 10 percent from $10.72 billion in 2006. Revenues for the Big Four climbed to $80 bilion from about $72 billion in 2006. The numbers exclude Shell Canada and Conoco

Phillips Canada, two private subsidiaries that produced almost 500,000 barrels per day in 2006.

- EnCana Corporation
- Canadian Natural Resources Limited
- Husky Energy Inc.
- Conoco Phillips
- Talisman Energy Inc.
- Devon Canada Corporation
- Suncor Energy
- Cenovus Energy.

Long-term Outlook

Broadly speaking Canadian oil production (via standard deep drilling) peaked in the mid-1970s, but due to the new offshore basins being exploited in Atlantic Canada and the boom of the Alberta Tar Sands overall production will not peak until the 2020s.

Agriculture

Canada is one of the largest agricultural producers and exporters in the world. As with other developed nations, the proportion of the population and GDP devoted to agriculture fell dramatically over the 20th century but it remains an important element of the Canadian economy.

A wide range of agriculture is practiced in Canada, from sprawling wheat fields of the prairies to wineries of the Okanagan valley. In the federal government, overview of Canadian agriculture is the responsibility of the department of Agriculture and Agri-Food.

History

In the 17th century Samuel de Champlain and Gabriel Sagard recorded that the Iroquois and Huron cultivated the soil for maize or "Indian corn". Maize (Zea mays), potatoes (Solanum tuberosum), beans (phaseolus), squash (Cucurbita) and the sunflower (Helianthus annus) were grown throughout agricultural lands in North America by the 16th century. As early as 2300 BC evidence of squash was introduced to the northeastern woodlands region. Archaeological findings from

500 AD have shown corn cultivation in southern Ontario. Eastern Canada was settled well before the West. Immigration and trading posts came later to Rupert's Land and the Northwest Territories. The early immigrants combined European agricultural and domestication procedures with the indigenous knowledge of the land and animals of the area.

As early as 1605, the French Acadians built dikes in the Maritimes for wheat, flax, vegetables, pasturage and marshland farming. Dairy production is the main contribution of New Brunswick, Nova Scotia, and Prince Edward Island, along with livestock and mixed farming ventures. A small percentage of land is put into use in fruit farming as well along Nova Scotia's northwest coastal areas. The American Revolution, 1775–1783, and its attendant food decline resulted in 3100 hectares cleared in Newfoundland. In the early 19th century Irish immigrants began arriving who cultivated the land in Newfoundland. A very small percentage of the land is suitable in Newfoundland and Labrador for horticultural or crop production because there is a lot of forested and tundra geography. The province has some dairy production and farming concerns. Following World War II, farm training was available at the Government Demonstration Farm. Bonuses were paid for such things as the purchase of pure-bred sires, land clearing, and agriculture exhibition assistance to name a few. The industry of fish processing for food is the largest agricultural contribution from Newfoundland. Newfoundland fisheries, supply cod for the most part, followed closely by herring, haddock, lobster, rose fish, seals, and whales. The fishing industry depends very heavily upon exports and world conditions.

Agriculture in the West started with Peter Pond gardening plots at Lake Athabasca in 1778. Although large-scale agriculture was still many years off, Hudson's Bay Company traders, gold rush miners, and missionaries cultivated crops, gardens and raised livestock. The Northwest Territories, Yukon, and Nunavut are covered with the Canadian Shield, and rocky outcrops, sub Arctic forest soils, and stony phases make up most of the geography. It is an area of comparatively smaller population and not commercially exploited for the most part. Whaling, prawns, and trapping food processing contribute to agricultural food production here.

In New France hops, hemp and livestock were introduced in 1663. The seigneurial system of farming was adopted in Quebec. Quebec's agricultural sector relies heavily on its fruit and vegetable production. In 1890, a competition began to encourage farmers to improve their farms to achieve the Agricultural Merit Order. County farm improvement contests were begun about 1930 involving over 5,000 farms and their evolution over five years. They have some interests in livestock and mixed farming and diary as well. St. Hyacinthe operated an artificial insemination station from 1951 for breeders clubs.

The British enforced Corn (Cereal grains) laws, 1794-1846, protected the British agricultural sector from imports of British North American wheat. The Reciprocity Treaty, June 6, 1854, developed a trade agreement between Canada and the United States which affected trade of wheat grown in Ontario. Northern Ontario is mainly tundra and forested area, whereas southern Ontario has lands suitable for livestock and general farming as well as geography suitable for pasture and dairying industries. Fruit farming and tobacco farms can also be found in southern Ontario. Ontario is the largest producer of mixed grains, soybeans and shelled corn in the country.

Lord Selkirk, founder of the Red River Colony, harvested the first wheat crop in the western prairies in 1814. Red Fife wheat was introduced in 1868. Swine were brought to the Red River colony as early as 1819. The frontier land of southwest Alberta and southeast Saskatchewan were opened to ranching in the 19th century. Manitoba has a combination of mixed grain, livestock, and mixed farming industries in its southernmost areas. Cattle ranching around Lake Manitoba is also quite successful. Northern Manitoba consists of extensive lakes and forested geographical areas. The Dominion Land Act of 1872 offered agricultural pioneers an opportunity to "prove up" a quarter section of land (160 acres/65 hectares]) in western Canada for a $10.00 filing fee and three years of improvements combined with residence on the land. Saskatchewan still has cattle ranching along its southwestern corner; grain farming and crops such as wheat, oats, flax, alfalfa, and rapeseed (especially canola) dominate the parkland area. Mixed grain farming, dairy farms, mixed livestock and grazing lands dot the central lowlands region of this prairie province.

Alberta is renowned still for its stampedes, and cattle ranching is a main industry. The agricultural industry is supplemented by livestock and mixed farming and wheat crops. Alberta is the second largest producer of wheat in Canada. Grain and dairying also play a role in the livelihoods of Alberta farmers.

The open parkland area extends across the three prairie provinces: Manitoba, Saskatchewan and Alberta. Canada's production of wheat, oats, flaxseed, and barley come mainly from this area. Meat processing is the largest industry here, followed by dairy production, breweries, and the subsidiary industry of agricultural implements.

British Columbia is covered in highlands; its eastern boundary is the Rocky Mountains. Livestock, cattle ranches, fruit farming and dairying dot the province. Agriculture and fisheries are a small contribution industry over shadowed by construction and forestry.

Agricultural production in British Columbia supplied the gold rush industry, mining and logging industries. Agricultural producers relied on these local markets, following the economic boom and bust of each enterprise respectively. The British Columbia Fruit-Growers' Association was established in 1889 to foster an export market of this commodity. The Canada Agriculture Museum preserves Canadian agricultural history.

Agricultural Museums:

- Canada Agriculture Museum
- Manitoba Agricultural Museum
- Ross Farm Museum
- Central Experimental Farm
- Ontario Agricultural Museum.

Crops

In 2008, Saskatchewan produced over half of the wheat in the Dominion of Canada, threshing in excess of over 9000 bushels (577000 metric tons) of wheat. North America has led other international continents as the main producer of wheat in total world production. Rapeseed, alfalfa, barley, canola, flax, rye, and oats are other popularly grown grain crops. Wheat is a staple crop from Canada. To help homesteaders attain an abundance harvest in a foreshortened growing season, varieties of wheat were developed at the beginning of the twentieth century. Red Fife was the first strain; it was a wheat which could be seeded in the fall and sprout in the early spring. Red Fife ripened nearly two weeks sooner and was a harder wheat than other spring wheats. Dr. C. Saunders, experimented further with Red Fife, and developed Marquis Wheat, which was resistant to rust and came to maturity within 100 days. Some other types of wheat grown are durum, spelt, and winter wheat. In recent years Canadian farmers have also began to grow rice due to the increase in the Asian population in Canada.

The Prairie Farm Recreational Administration was established in 2008 to provide Federal financial assistance in regard to the global economical crisis. The Prairie Farm Recreational Administration provides farmers with land and water resources such as irrigation, soil drifting conservation and small farm water development. The Farm credit program has established the Canadian Farm Loan Act to provide stock bonds and farm improvement loans.

Horticulture

Horticulture which includes garden crops, and fruits became easier to grow with the development of plant hardiness zones. Apples, pears, plums and prunes, peaches, apricots, cherries, strawberries, raspberries, loganberries and fruit orchards are numerous and reach commercial size in the Annapolis Valley of Nova Scotia, New Brunswick, Quebec, Niagara Peninsula and Norfolk County of Ontario and Okanagan Valley of British Columbia. Hazelnuts are harvested in Eastern Canada and British Columbia. Maple syrup and maple sugar, maple butter, and maple taffy are products of Quebec along the St. Lawrence River. The main market for Canadian maple syrup and sugar is the United States Potatoes are an abundant harvest of the Maritime provinces. Tobacco is an agricultural commodity from the Ontario tobacco belt, in particular Norfolk County, adjacent to Lake Erie and Quebec. Sugar beets and beet root sugar are harvested in Quebec, Ontario, Manitoba, and Alberta.

Viticulture

Viticulture refers to the growing of grapes. Grapes require a mild winter season, which can be found in some Maritime locations, southern British Columbia, and locations on the Niagara Peninsula.

Livestock

115,000 cattle roamed the southern prairies by 1900. Livestock can include the raising of cows, also commonly called cattle. Recently domestication of the buffalo and elk has initiated a new food industry. Sheep have been raised for both wool and meat. Bovine or pig barns have been a part of livestock culture. Scientists have been making forward steps in swine research giving rise to intensive pig farming. The domestication of various farm animals meant that corresponding industries such as feedlots, animal husbandry and meat processing have also been studied, and developed.

Dairy, Poultry and Eggs

Fowl, poultry, eggs, chickens, geese, ducks and turkeys are part of a supply-managed system, ensuring production matches demand. Dairy producing is also termed dairy farming. Butter production in Canada was on average 330,000,000 pounds (150,000 t) in the 1940s; and cheese production between 95,000,000 pounds and 208,000,000 pounds (43,000 t to 94,000 t) in that same time. The United Kingdom received 50,000,000 pounds (23,000 t) in 1949.

The Canadian Dairy Commission Act was passed in 1966 by the Canadian federal Government to provide efficient producers of milk and cream with the opportunity of obtaining a fair return for their labour and investment and to provide consumers with a continuous and adequate supply of dairy products of high quality.– Canadian Dairy Commission

In the 1970s the supply management system came into effect to regulate supply of milk, poultry and egg to meet consumer demand. The collective marketing ensures that imports are limited in areas where product can be supplied domestically. The federal government imposed pricing policies to safeguard the producer's livelihood.

Other

In recent years farmers have been producing alternative crops which are economically viable, and amongst these are organic farm crops. Hemp and wool from sheep are the main areas of fiber production of Canada. Wool production was on average 16,022,000 pounds (7,267 t) in the 1930s and 9,835,000 pounds] (4,461 t) in 1949. Fibre flax from flaxseed has been exported to the United Kingdom. Crop growers may supplement their income with beeswax and honey and learn beekeeping which is overseen by the apiary branch. Enterprising land owners have had success growing as well as packaging and marketing the sunflower seed. Crops are not only for human consumption but also for animal consumption, which opens a new market such as canary seed. Cuniculture, or rabbit farming are a new grocery alternative to the red meat burger. Cannabis is an important crop in some areas, making up 5% of British Columbia's GDP. According to BC Business Magazine, the crop is worth $7.5 billion to the province annually, and gives employment to 250,000 people. Quebec produces an even bigger crop.

Number of farms by Province/Territory

Province / Territory	*Amount of Farms*
Alberta	53,652
British Columbia	20,290
Manitoba	21,071
New Brunswick	3,034
Newfoundland and Labrador	643
Nova Scotia	3,923
Ontario	59,729
Prince Edward Island	1,845
Quebec	32,139
Saskatchewan	50,598
Northwest Territories	30
Nunavut	0
Yukon	170

Canadian Agricultural Government Departments

The Department of Agriculture set out in the British North America Act (B.N.A.) of 1867 states each province may have jurisdiction over agricultural concerns, as well as the Dominion Government may also make law in regard to agriculture. Newfoundland agricultural affairs were dealt with by the

Agricultural Division of the Department of Natural Resources at Confederation.

The B.N.A. Act states that the federal Government has sole authority in coastal and inland fishery matters. Provinces have rts over non-tidal waters and fishing practices there only.

Canadian agricultural government departments

Department	***Function***
Agriculture and Agri-Food Canada	Responsible for policies governing agriculture production, farming income, research and development, inspection, and the regulation of animals and plants. Headed by the Minister of Agriculture (Canada).
Canadian Dairy Commission	Responsible for providing dairy producers a fair return for labour and investment and provide consumers with high quality dairy products.
Canadian Food Inspection Agency	CFIA consolidates the delivery of all federal food, animal and plant health inspection programs.
Canadian Grain Commission Canadian Wheat Board	Responsible for the grain industry Headed by the Minister of Agriculture and Agri-food
Fisheries and Oceans Canada	Responsibility for the conservation and sustainable use of Canada's fisheries resources.
National Farm Products Council	Responsible for promoting efficient and competitive agriculture in Canada and oversees the Canadian Egg Marketing Agency, Canadian Turkey Marketing Agency, Chicken Farmers of Canada and Canada Hatching Egg Producers.

Agricultural Economy

Canadian farms, fisheries and ranches produce a wide variety of crops, livestock, food, feed, fiber, fuel and other goods by the systematic raising of plants and animals which are dependent upon the geography of the province. In 2001 farms numbered only 246,923 at a size of 676 acres (2.74 km^2) as the production of food and fiber for human or livestock sustenance has evolved into intensive and industrial practices. As of 2002, wheat constituted the largest crop area at 12.6%. Canadian farmers received a record $36.3 billion in 2001 from livestock, crop sales and program payments. In 2001, the accrued net income of farm operators from farm production amounted to 1,633 million dollars, which amounts to 0.147% of Canada's gross domestic product at market prices which is 1,108,200 million dollars. Fisheries are also playing an important role while forestry plays a secondary role. Canada's evolution has abandoned subsistence techniques and now sees a mere 3% of Canada's population employed as a mechanized industrial farmer who are able feed the rest of the nation's population of 30,689.0 thousand people (2001) as well as export to foreign markets. (Canada's estimated population was 32,777,300 on January 1, 2007).

Trade

The marketing and economic movement of Canada's various agriculture commodities has been a challenge. Domestic trade encompasses providing goods within Canada provincially and inter-provincial. Support agencies and services such as storage, railways, warehouses, stores, banking institutions all effect domestic trade. Trade of wheat from the Canada's prairies are monitored by the Canadian Wheat Board. Canada's depression of 1882-1897 brought a low of 64¼ cents per bushel ($24/t) as of 1893. This era during Laurier's administration saw thousands of homesteads cancelled. Wheat prices soared during World War I. In 1928, Canada exported high quantities of wheat, flour, and goods. The depression took its toll on Canada as exports sunk to approximately 40% of their 1928 amount. European markets stopped needing to import Canadian wheat as they started growing their own varieties, and then World War II events put a blockade on trade to European markets. Canada became more of an industrial entity during the time of this industrial revolution, and less of an agricultural nation. Following World War II the United Kingdom entered into contract for a large amount of agricultural commodities such as bacon, cheese, wheat, oats and barley. After the United Kingdom, the United States is Canada's largest external trade partner. Between 1943 and 1953, the average export of Canadian wheat was 347,200,000 bushels (9,449,000 t). The three year International Wheat Agreement of 1955, included exports of wheat or flour to 28 of 44 importing countries including Germany, Japan, Belgium, UK, and the Netherlands.

Agribusiness

Agribusiness are activities of food and fibre production and processing which are not part of the farm operation. This would include the production of farm equipment and fertilizers to aid farm production. Agribusiness also includes the firms that purchase the raw goods from the farm for further processing. The meat packing industry, flour mill, and canning industry would be included in the agribusiness sector processing farm products.

Industry Categories

According to Agriculture and Food Canada, these are the classifications of Canadian Agriculture Industries.

Canadian Agriculture Industries

Industry	*Mainstay*
Brewery industry	Comprises two large national beer producing companies: Labatt Breweries of Canada and Molson Canada Breweries
Buckwheat industry	Buckwheat flour is used for pancake mixes and pasta. Buckwheat is exported mainly to Japan. Majority of this specialty crop grown in Manitoba
Canary seed industry	In 2005, Canada produced 77% of the world canary seed production. Saskatchewan soils were conducive to bird seed production.
Confectionery andchewing gum industry	Sugar and cocoa are imported for this industry which has foreign owned firms operating in Canada. Various candies amounting to $1.48 billion were shipped in 1997.
Dairy industry	In the Canadian agri-food economy the dairy industry is the third largest.

Dairy genetics industry	The Canadian Record of Performance R.O.P. program discovers dairy cattle of high producing milk capacities. Cattle qualities are monitored by the Canadian Dairy Herd Improvement milk producing agency. (Canadian DHI).
Distillery industry	Canadian whisky made from rye and corn is the main aspect of this Canadian industry. The distillery industry also includes production of whisky, rum, vodka, gin, liqueurs, spirit coolers and basic ethyl alcohol.
Egg industry	Evolved into an automated industry producing table eggs, enzymes, breaker eggs, processed foods, and supporting pullet producers, egg laying chicken (layers) producers and graders.
Fish and seafood industry	This industry produces CDN $5 billion a year. The world's fourth-largest exporter of fish is Canada, from the Atlantic fishery, Pacific fishery and aquaculture sector.
Forage industry	This industry comprises feed for livestock, cattle, sheep and horses. Hay is the main forage crop, supplemented by alfalfa, cereals, peas and corn. Besides domestic markets, exports from Canada arrive at Pacific Rim Countries.
Fruit industry	Tree fruit grower crops consist of apples, pears, peaches, plums, apricots, nectarines, and sweet cherry, followed by wine grape areas. The industry supports fresh, canned, frozen and preserved fruits as well as food production.
Grains and oilseeds industry	Wheat, barley and oats are Canada's grain exports. Canola, soybean and flaxseed are the main oilseed exports.
Grain-based products industry	Grain and oilseed production supports flour milling, malt manufacturing, starch, vegetable fat and oil manufacturing as well as breakfast cereal manufacturing

Hemp industry	Spin off industries from Hemp production include aromatherapy, commercial oil paints, cosmetics, edible oil, garments and accessories, hemp meal and flour, snack foods, shampoo and conditioners, and moisturizers.
Honey industry	Beeswax produces cosmetics, ointments, candles and household waxes. A diet supplement is made from bee pollen. Propolis and royal jelly is used in cosmetics, creams, lotions, tonics and lip balms. Honey is a sweetener for domestic use or commercial food production.
Industrial agriculture (animals)	Factory farming, Intensive pig farming, Integrated Multi-trophic Aquaculture, and shrimp farming are various forms of industrial agriculture which aims at mass production
Industrial agriculture	Includes innovation in agricultural machinery and farming methods, genetic technology, techniques for achieving economies of scale in production, the creation of new markets for consumption, the application of patent protection to genetic information, and global trade
Maple syrup industry	Maple syrup can be used to make maple sugar, maple butter, maple taffy as well as a sweetener.
Mustard seed industry	Yellow mustard is the highest export, closely followed by brown and oriental mustards. 2007 saw an increase in mustard seed prices.
Organic industry	Operational certification and standards are challenges for the growing organic farming industry. Organic farming with biodynamics and without synthetic chemicals provides the consumer a holistic plant and animal food choice.

Potato industry	Potato Innovation Network (PIN) 2020 was initiated in 2006 to support development of new markets, and new uses for potatoes in market diversification.
Poultry industry	Avian Influenza ("Bird Flu") is the latest concern in the poultry industry however disease precautions are in place if this strain arrives in Canada.
Processed fruit and vegetable industry	Processing of fruits and vegetables includes consumer products of canned, cider, frozen, jams, jellies and marmalades, pickles, sauces, soups, vegetable and fruit juices and vinegar.
Pulse industry	Beans, chickpeas, faba beans, and lentils comprise the pulse industry. Peas soup, and baked beans are large production processes from pulse growth. The world's largest pulse exporter is Canada.
Red meat industry	This is Canada's fourth major manufacturing industry. Cattle, calves, hogs, sheep, lambs, venison, bison are all domesticated for red meat export and domestic consumption.
Seed industry	Seed growers, field inspectors, registered seed establishments, seed trial plots, and seed retailers are the mainstays of seed production.
Snack food industry	Cereal grains, cornmeal, nuts, oils, potatoes, and seeds are the major ingredients of snack foods such as potato chips, mixed nuts, peanut butter, pork rinds, and seed snacks.
Sunflower seed industry	About 80 per cent sunflowers grown in Canada are sold as roasted snack sunflower seeds or without the shell for baking. The main consumer is domestic. Birdfeed and sunflower vegetable oils are smaller markets which are being developed.

Vegetable industry	The edible portion of a plant is a vegetable. Vegetables can be marketed fresh or as part of the processed fruit and vegetable industry. The greenhouse vegetable industry supports the field vegetable farmer.
Wine industry	Canadian vintners producing wines with unique aromas, aging characteristics and flavours bring in international awards. The grape hybrid from the native Canadian species bred with wine producing grapes results in a grape for a shorter, cooler growing season, and a quality not found elsewhere.

Agricultural Science

Agricultural science began developing new styles of farming and strains of wheat and crops so that farming could become a successful venture. Farming methods were developed at places such as Indian Head Experimental Farm, Rosthern Experimental Station, and Bell Farm. The Better Farming Train travelled around rural areas educating pioneer farmers. The 1901 census showed 511,100 farms and the number of farms peaked in 1941 at a record 732,800 farms. The industrial revolution modernized the farming industry as mechanized vehicles replaced the oxen ploughed land or the horse drawn cart. Farms became much larger, and mechanized evolving towards industrial agriculture.

Production

Farming activities were very labour intensive before the industrial revolution and the advent of tractors, combines, balers, etc. In the late 1800s to mid 1900s, a great percentage of the Canadian labour force was engaged in high labour, smaller farming practices.

After mechanization, scientific advancement, improved marketing practices farms became more efficient, larger and less labour intensive. The labour population was freed up and went to industry, government, transportation, trade and finance. Agriculture, stock raising and horticulture employed one-fourth of the Canadian population according to the 1951 census as well as providing products for exports and Canadian manufacturing concerns.

Farm Equipment

The Oliver plow was in use by 1896 which could cut through the prairie sod. Binders which could cut and tie grain for the harvest season and grain elevators for storage were introduced in the late 1800s as well. Plows, tractors, spreaders, combines to name a few are some mechanized implements for the grain crop or horticultural farmer which are labour saving devices. Many Canadian museums such as Reynolds-Alberta Museum will showcase the evolution and variety of farm machinery.

Challenges

The depression and drought of the Dirty Thirties was devastating. This drought resulted in a mass exodus of population from the prairies, as well as new agricultural practices such as soil conservation, and crop rotation.

Soil conservation practices such as crop rotation, cover crops, and windbreaks to name a few were massively developed and set in forth upon recovering from the drought experiences of the dirty thirties. Literally layers and layers of topsoil would be blowing away during this time. Bow River Irrigation Project, Red Deer River Project and the St. Mary Irrigation project of Alberta, were a few of the major projects undertaken by the Prairie Farm Rehabilitation Act (P.F.R.A.) resulting in reservoirs, and distribution systems. A current project is Liming (soil) soil liming at the Land Resource Research Institute. Wheat diseases such as wheat bunt and stinking smut can be successfully treated with a fungicide. Disease of plants and animals can break an agricultural producer. Tuberculosis in animals was an early threat, and cattle needed to be tested, and areas accredited in 1956. The newer disease such as chronic wasting disease or transmissible spongiform encephalopathy (TSE) affects both elk and deer. Elk and deer raising is a pioneer field of domestication, has had a setback with this disease. Mad cow disease in cattle and scrapie of sheep are monitored by the Canadian Food Inspection Agency. The poultry sector was plagued by Pullorum disease, and by controlling the flock via poultry husbandry, this disease has been brought under control.

Plants whose traits can be modified to survive a disease or insect have made inroads into Canadian agricultural practices. Cereal rusts which can destroy the majority of areas seeded to wheat, was controlled in 1938 by breeding strains which were rust-resistant.

This strain was successful until around 1950, when again a new variety of rust broke out, and again a new species of wheat called Selkirk was developed which was rust resistant. Biotechnology is the centre of new research and regulations affecting agriculture this century.

Developmental and Educational Institutions

To increase the viability of agriculture as an economic lifestyle several improvements have been made by various nationwide educational facilities. Inroads and innovations have been made in the diverse fields of agricultural science, agricultural engineering, agricultural soil science, Sustainable agriculture, Agricultural productivity, agronomy, biodiversity, bioengineering, irrigation and swine research for example.

Canada is also one of the world's largest suppliers of agricultural products, particularly of wheat and other grains. Canada is a major exporter of agricultural products, to the United States but also to Europe and East Asia. As with all other developed nations the proportion of the population and GDP devoted to agriculture fell dramatically over the 20th century.

As with other developed nations, the Canadian agriculture industry receives significant government subsidies and supports. However, Canada has been a strong supporter of reducing market influencing subsidies through the World Trade Organization.

In 2000, Canada spent approximately CDN$4.6 billion on supports for the industry. Of this, $2.32 billion was classified under the WTO designation of "green box" support, meaning it did not directly influence the market, such as money for research or disaster relief. All but $848.2 million were subsidies worth less than 5% of the value of the crops they were provided for, which is the WTO threshold. Consequently, Canada used only $848.2 million of its $4.3 billion subsidy allowance granted by the WTO.

Manufacturing

The general pattern of development for wealthy nations was a transition from a primary industry based economy to a manufacturing based one, and then to a service based economy. Canada did not escape this pattern-at it's (abnormally high World War 2) peak in 1944, manufacturing accounted for 29% of GDP, declining to 15.6% in 2005. Canada has not suffered as greatly as most other rich, industrialized nations from the pains of the relative decline in the importance of manufacturing since the 1960's. A 2009 study by Statistics Canada also found that, while manufacturing declined as a relative percentage of GDP from 24.3% in the 1960's to 15.6% in 2005, manufacturing volumes between 1961 and 2005 kept pace with the overall growth in the volume index of GDP. Manufacturing in Canada was especially hard hit by the 2007-2010 global financial crisis. As of 2010, manufacturing accounts for 13% of Canada's GDP, a relative decline of more than 2% of GDP since 2005.

Central Canada is home to branch plants to all the major American and Japanese automobile makers and many parts factories owned by Canadian firms such as Magna International and Linamar Corporation. Central Canada today produces more vehicles each year than the neighbouring U.S. state of Michigan, the heart of the American automobile industry. Manufacturers have been attracted to Canada due to the highly educated population with lower labour costs than the United States. Canada's publicly funded health care system is also an important attraction, as it exempts companies from the high health insurance costs they must pay in the United States.

Much of the Canadian manufacturing industry consists of branch plants of United States firms, though there are some important domestic manufacturers, such as Bombardier Inc.. This has raised several concerns for Canadians. Branch plants provide mainly blue collar jobs, with research and executive positions confined to the United States.

Service Sector

The service sector in Canada is vast and multifaceted, employing some three quarters of Canadians and accounting for over two thirds of GDP. The largest employer is the retail sector, employing almost 12% of Canadians. The retail industry is mainly concentrated in a relatively small number of chain stores clustered together in shopping malls. In recent years the rise of big-box stores, such as Wal-Mart (of the United States) and Future Shop (a subsidiary of the US based Best Buy and Wal-Mart and Zellers. This has led to fewer workers in this sector and a migration of retail jobs to the suburbs.

The second largest portion of the service sector is the business services, employing only a slightly smaller percentage of the population. This includes the financial services, real estate, and communications industries. This portion of the economy has been rapidly growing in recent years. It is largely concentrated in the major urban centres, especially Toronto, Montreal and Vancouver.

The education and health sectors are two of Canada's largest, but both are largely under the purview of the government. The health care industry has been rapidly growing, and is the third largest in Canada. Its rapid growth has led to problems for governments who must find money to fund it.

Canada has an important high tech industry, and also an entertainment industry creating content both for local and international consumption. Tourism is of ever increasing importance, with the vast majority of international visitors coming from the United States. Though the recent strength of the Canadian Dollar has hurt this sector, other nations such as China have increased tourism to Canada.

Political Issues

Regional Imbalances

The Canadian economy differs greatly from region to region. Traditionally Central Canada has been the economic engine of Canada, home to more than half of its population and much of its industry. Recent years have seen rapid growth in Western Canada as trade with Asia has enriched British Columbia and oil wealth provided a major boost to Alberta and Saskatchewan.

The four Atlantic provinces, though once the centre of economic activity, underwent a major decline in the late 19th century and have traditionally been significantly poorer than the rest of Canada, especially after the recent collapse of the fishing industry. Recent years have seen some significant moves towards diversification, especially as offshore oil and gas wealth have begun to flow into the region. Quebec has also traditionally been poorer than the Canadian average although by a lesser margin than the Atlantic provinces. In more recent years Newfoundland and Labrador have started to see a change in their economy, being called the "Celtic tiger of Canada," (in a comparison to the economic transformation in Ireland); it has also been called a "mini Alberta" because of new oil and gas exploration. While many Newfoundlanders and Labradorians still emigrate to Alberta for higher-paying jobs, the provinces population has been on the rise in recent years with many people deciding to move back home.

Relations with the U.S.

Canada and the United States share a deep and common trading relationship. Canada's job market continues to perform well along with the US, reaching a 30 year low in the unemployment rate in December 2006, following 14 consecutive years of employment growth.

The United States is by far Canada's largest trading partner, with more than $1.7 billion CAD in trade per day in 2005. In 2009 73% of Canada's exports go to the United States, and 63% of Canada's imports are from the United States. Trade with Canada makes up 23% of America's exports and 17% of its imports. By comparison, in 2005 this was more than U.S. trade with all countries in the European Union combined, and well over twice U.S. trade with all the countries of Latin America combined. Just the two-way trade that crosses the Ambassador Bridge between Michigan and Ontario equals all U.S. exports to Japan. Canada's importance to the United States is not just a border-state phenomenon: Canada is the leading export market for 35 of 50 U.S. states, and is the United States' largest foreign supplier of energy.

Bilateral trade increased by 52% between 1989, when the U.S.-Canada Free Trade Agreement (FTA) went into effect, and 1994, when the North American Free Trade Agreement (NAFTA) superseded it. Trade has since increased by 40%. NAFTA continues the FTA's moves toward reducing trade barriers and establishing agreed upon trade rules. It also resolves some long-standing bilateral irritants and liberalizes rules in several areas, including agriculture, services, energy, financial services, investment, and government procurement. NAFTA forms the largest trading area in the world, embracing the 405 million people of the three North American countries.

The largest component of U.S.-Canada trade is in the commodity sector. The U.S. is Canada's largest agricultural export market, taking well over half of all Canadian food exports. Similarly, Canada is the largest market for U.S. agricultural goods with nearly 20% of American food exports going to its Northern neighbor. Nearly two-thirds of Canada's forest products, including pulp and paper, are exported to the United States; 72% of Canada's total newsprint production also is exported to the U.S. At $73.6 billion in 2004, U.S.-Canada trade in energy is the largest U.S. energy trading relationship, with the overwhelming majority ($66.7 billion) being exports from Canada. The primary components of U.S. energy trade with Canada are petroleum, natural gas, and electricity. Canada is the United States' largest oil supplier and the fifth-largest energy producing country in the world. Canada provides about 16% of U.S. oil imports and 14% of total U.S. consumption of natural gas. The United States and Canada's national electricity grids are linked and both countries share hydro power facilities on the Western borders.

While most of U.S.-Canada trade flows smoothly, there are occasionally bilateral trade disputes, particularly in the agricultural and cultural fields. Usually these issues are resolved through bilateral consultative forums or referral to World Trade Organization (WTO) or NAFTA dispute resolution. In May 1999, the U.S. and Canadian Governments negotiated an agreement on magazines that provides increased access for the U.S. publishing industry to the Canadian market. The United States and Canada also have resolved several major issues involving fisheries. By common agreement, the two countries submitted a Gulf of Maine boundary dispute to the International Court of Justice in 1981; both accepted the Court's 12 October 1984 ruling which demarcated the territorial sea boundary. A current issue between the United States and Canada is the ongoing softwood lumber dispute, as the U.S. alleges that Canada unfairly subsidizes its forestry industry. In 1990, the United States and Canada signed a bilateral Fisheries

Enforcement Agreement, which has served to deter illegal fishing activity and reduce the risk of injury during fisheries enforcement incidents. The U.S. and Canada signed a Pacific Salmon Agreement in June 1999 that settled differences over implementation of the 1985 Pacific Salmon Treaty for the next decade. Canada and the United States signed an aviation agreement during Bill Clinton's visit to Canada in February 1995, and air traffic between the two countries has increased dramatically as a result. The two countries also share in operation of the St. Lawrence Seaway, connecting the Great Lakes to the Atlantic Ocean.

The U.S. is Canada's largest foreign investor and the most popular destination for Canadian foreign investments; at the end of 2007, the stock of U.S. direct investment in Canada was estimated at $293 billion while Canadian direct investment (stock) in the United States was valued at $213 billion. U.S FDI accounts for 59.5% of total foreign direct investment in Canada while Canadian F.D.I in U.S accounts for 10% (5th largest foreign investor). US investments are primarily directed at Canada's mining and smelting industries, petroleum, chemicals, the manufacture of machinery and transportation equipment, and finance while Canadian investment in the United States is concentrated in manufacturing, wholesale trade, real estate, petroleum, finance, and insurance and other services

Free Trade Agreements:

- Canada-U.S. Free Trade Agreement (Signed 12-Oct-1987, entered into force 01-Jan-1989, later superseded by NAFTA)
- North American Free Trade Agreement (Entered into force 01-Jan-1994, includes Canada, U.S. and Mexico)
- Canada-Israel Free Trade Agreement (Entered into force 01-Jan-1997)
- Canada-Chile Free Trade Agreement (Entered into force 05-Jul-1997)
- Canada-Costa Rica Free Trade Agreement (Entered into force 01-Nov-2002)
- Canada-European Free Trade Association Free Trade (EFTA) Agreement (Entered into force 01-Jul-2009)

- Canada-Peru Free Trade Agreement (Entered into force 01-Aug-2009)
- Canada-Colombia Free Trade Agreement (Entered into force 21-Nov-2008)
- Canada-Jordan Free Trade Agreement (Entered into force 28-June-2009)
- Canada-Panama Free Trade Agreement (Entered into force 14-May-2010).

Canada is negotiating bilateral FTAs with the following countries and trade blocs:

- Ukraine
- Morocco
- India
- South Korea
- Dominican Republic
- Singapore
- Andean Community (Negotiations have already concluded with Peru and Colombia)
- CARICOM (Caribbean Community)
- European Union.

Canada is also involved in negotiations to create the following regional trade blocks:

- Canada Central American Free Trade Agreement
- Free Trade Area of the Americas (FTAA).

Canada is one of the world's wealthiest nations, with a high per-capita income, and it is a member of the Organisation for Economic Co-operation and Development (OECD) and the G8. It is one of the world's top ten trading nations. Canada is a mixed market, ranking above the U.S. on the Heritage Foundation's index of economic freedom and higher than most western European nations. The largest foreign importers of Canadian goods are the United States, the United Kingdom, and Japan. In 2008, Canada's imported goods were worth over $442.9 billion, of which $280.8 billion was from the United States, $11.7 billion from Japan, and $11.3 billion from the United Kingdom. The country's 2009 trade deficit totaled C$4.8 billion, compared with a C$46.9 billion surplus in 2008.

As of October 2009, Canada's national unemployment rate was 8.6%. Provincial unemployment rates vary from a low of 5.8% in Manitoba to a high of 17% in Newfoundland and Labrador. Canada's federal debt is estimated to be $566.7 billion for 2010-11, up from $463.7 billion in 2008-09. Canada's net foreign debt rose by $40.6-billion to $193.8-billion in the first quarter of 2010. The combined federal and provincial government deficit in the 2009-10 fiscal year could reach of $100-billion, and the federal deficit is forecast to be C$49.2 billion in 2010-11.

In the past century, the growth of the manufacturing, mining, and service sectors has transformed the nation from a largely rural economy to a more industrial and urban one. Like other First World nations, the Canadian economy is dominated by the service industry, which employs about three quarters of Canadians. Canada is unusual among developed countries in the importance of its primary sector, in which the logging and petroleum industries are two of the most important.

Canada is one of the few developed nations that are net exporters of energy. Atlantic Canada has vast offshore deposits of natural gas, and Alberta has large oil and gas resources. The immense Athabasca Oil Sands give Canada the world's second-largest oil reserves, behind Saudi Arabia.

Canada is one of the world's largest suppliers of agricultural products; the Canadian Prairies are one of the most important producers of wheat, canola, and other grains. Canada is the largest producer of zinc and uranium, and is a global source of many other natural resources, such as gold, nickel, aluminium, and lead. Many towns in northern Canada, where agriculture is difficult, are sustainable because of nearby mines or sources of timber. Canada also has a sizable manufacturing sector centred in southern Ontario and Quebec, with automobiles and aeronautics representing particularly important industries.

Economic integration with the United States has increased significantly since World War II. This has drawn the attention of Canadian nationalists, who are concerned about cultural and economic autonomy in an age of globalization, as American goods and media products have become ubiquitous. The Automotive Products Trade Agreement of 1965 opened the borders to trade in the auto manufacturing industry. In the 1970s, concerns over energy self-sufficiency and foreign ownership in the manufacturing sectors prompted Prime Minister Pierre Trudeau's Liberal government to enact the National Energy Program (NEP) and the Foreign Investment Review Agency (FIRA). In the 1980s, Prime Minister Brian Mulroney's Progressive Conservatives abolished the NEP and changed the name of FIRA to "Investment Canada" in order to encourage foreign investment. The Canada – United States Free Trade Agreement (FTA) of 1988 eliminated tariffs between the two countries, while the North American Free Trade Agreement (NAFTA) expanded the free-trade zone to include Mexico in the 1990s. In the mid-1990s, the Liberal government under Jean Chretien began to post annual budgetary surpluses and steadily paid down the national debt. The 2008 global financial crisis caused a recession, which could boost the country's unemployment rate to 10%.

Demographics

Canada's 2006 census counted a total population of 31,612,897, an increase of 5.4% since 2001. Population growth is from immigration and, to a lesser extent, natural growth. About four-fifths of Canada's population lives within 150 kilometres (93 mi) of the United States border. A similar proportion live in urban areas concentrated in the Quebec City – Windsor Corridor (notably the Greater Golden Horseshoe, including Toronto and area, Montreal, and Ottawa), the BC Lower Mainland (consisting of the region surrounding Vancouver), and the Calgary–Edmonton Corridor in Alberta.

According to the 2006 census, the largest reported ethnic origin is English (21%), followed by French (15.8%), Scottish (15.2%), Irish (13.9%), German (10.2%), Italian (5%), Chinese (3.9%), Ukrainian (3.6%), and First Nations (3.5%). Approximately one third of respondents identified their ethnicity as "Canadian". There are 600 recognized First Nations governments or bands encompassing 1,172,790 people. Canada's Aboriginal population is growing at almost twice the national rate, and 3.8% of Canada's population claimed aboriginal identity in 2006. Another 16.2% of the population belonged to non-aboriginal visible minorities. The largest visible minority groups in Canada are South Asian (4%), Chinese (3.9%) and Black (2.5%). In 1961, less than 2% of Canada's population (about 300,000 people) could be classified as belonging to a visible minority group and less than 1% as aboriginal. In 2006, 51.0% of Vancouver's population and 46.9% of Toronto's population were members of visible minority groups. Between 2001 and 2006, the visible minority population rose by 27.2%. According to a 2005 forecast by Statistics Canada, the proportion of Canadians belonging to a visible minority group in Canada could reach as much as 23% by 2017. As of 2007, almost one in five Canadians (19.8%) were foreign-born. Nearly 60% of new immigrants hail from Asia (including the Middle East). By 2031, one in three Canadians could belong to a visible minority group. Canada has the highest per-capita immigration rate in the world, driven by economic policy and family reunification, and is aiming for between 240,000 and 265,000 new permanent residents in 2010. Canada also accepts large numbers of refugees. New immigrants settle mostly in major urban areas like Toronto and Vancouver.

In common with many other developed countries, Canada is experiencing a demographic shift towards an older population, with more retirees and fewer people of working age. In 2006, the average age of the population was 39.5 years. The census results also indicate that despite an increase in immigration since 2001 (which gave Canada a higher rate of population growth than in the previous intercensal period), the aging of Canada's population did not slow during the period.

Support for religious pluralism is an important part of Canada's political culture. According to the 2001 census, 77.1% of Canadians identify as being Christians; of this, Catholics make up the largest group (43.6% of Canadians). The largest Protestant denomination is the United Church of Canada (9.5% of Canadians), followed by the Anglicans (6.8%), Baptists (2.4%), Lutherans (2%), and other Christians (4.4%). About 16.5% of Canadians declare no religious affiliation, and the remaining 6.3% are affiliated with non-Christian religions, the largest of which is Islam (2.0%), followed by Judaism (1.1%). Canadian provinces and territories are responsible for education. Each system is similar, while reflecting regional history, culture and geography. The mandatory school age ranges between 5–7 to

16–18 years, contributing to an adult literacy rate of 99%. Post-secondary education is also administered by provincial and territorial governments, which provide most of the funding; the federal government administers additional research grants, student loans, and scholarships. In 2002, 43% of Canadians aged 25 to 64 possessed a post-secondary education; for those aged 25 to 34, the rate of post-secondary education reached 51%.

Culture

Canadian culture has historically been influenced by British, French, and aboriginal cultures and traditions. There are distinctive Aboriginal cultures, languages, art, and music spread across Canada. Many North American Indigenous words, inventions and games have become an everyday part of Canadian language and use. The canoe, snowshoes, the toboggan, lacrosse, tug of war, maple syrup and tobacco are examples of products, inventions and games. Some of the words include the barbecue, caribou, chipmunk, woodchuck, hammock, skunk, mahogany, hurricane and moose. Numerous areas, towns, cities and rivers of the Americas have names of Indigenous origin. The province of Saskatchewan derives its name from the Cree language name of the Saskatchewan River, "Kisiskatchewani Sipi". Canada's capital city Ottawa comes from the Algonquin language term "adawe" meaning "to trade." National Aboriginal Day recognises the cultures and contributions of Aboriginal peoples of Canada.

Canadian culture has been greatly influenced by immigration from all over the world. Many Canadians value multiculturalism and see Canada as being inherently multicultural. However, the country's culture has been heavily influenced by American culture because of its proximity and the high rate of migration between the two countries. The great majority of English-speaking immigrants to Canada between 1755 and 1815 were Americans from the Thirteen Colonies; during and immediately after the American Revolutionary War, 46,000 Americans loyal to the British crown came to Canada. Between 1785 and 1812, more Americans emigrated to Canada in response to promises of land. American media and entertainment are popular, if not dominant, in English Canada;

conversely, many Canadian cultural products and entertainers are successful in the United States and worldwide. Many cultural products are marketed toward a unified "North American" or global market. The creation and preservation of distinctly Canadian culture are supported by federal government programs, laws, and institutions such as the Canadian Broadcasting Corporation (CBC), the National Film Board of Canada, and the Canadian Radio-television and Telecommunications Commission.

Canadian visual art has been dominated by Tom Thomson — Canada's most famous painter — and by the Group of Seven. Thomson's brief career painting Canadian landscapes spanned just a decade up to his death in 1917 at age 39. The Group were painters with a nationalistic and idealistic focus, who first exhibited their distinctive works in May 1920. Though referred to as having seven members, five artists — Lawren Harris, A. Y. Jackson, Arthur Lismer, J. E. H. MacDonald, and Frederick Varley — were responsible for articulating the Group's ideas. They were joined briefly by Frank Johnston, and by commercial artist Franklin Carmichael. A. J. Casson became part of the Group in 1926. Associated with the Group was another prominent Canadian artist, Emily Carr, known for her landscapes and portrayals of the indigenous peoples of the Pacific Northwest Coast.

Canada has developed a music infrastructure and industry, with broadcasting regulated by the Canadian Radio-television and Telecommunications Commission. The Canadian music industry has produced internationally renowned composers, musicians and ensembles, such as Portia White, Guy Lombardo, Murray Adaskin, Rush, Joni Mitchell and Neil Young. Canadian winners of multiple Grammy Awards have included Celine Dion, k.d. lang, Sarah McLachlan, Alanis Morissette and Shania Twain. The Canadian Academy of Recording Arts and Sciences administers Canada's music industry awards, the Juno Awards, which commenced in 1970. The national anthem of Canada O Canada adopted in 1980, was originally commissioned by the Lieutenant Governor of Quebec, the Honourable Theodore Robitaille, for the 1880 St. Jean-Baptiste Day ceremony. Calixa Lavallee wrote the music, which was a setting of a patriotic poem composed by the poet and judge Sir Adolphe-Basile

Routhier. The text was originally only in French, before it was translated to English in 1906.

Canada's National symbols are influenced by natural, historical, and Aboriginal sources. The use of the maple leaf as a Canadian symbol dates to the early 18th century. The maple leaf is depicted on Canada's current and previous flags, on the penny, and on the Coat of Arms. Other prominent symbols include the beaver, Canada Goose, Common Loon, the Crown, the Royal Canadian Mounted Police, and more recently, the totem pole and Inukshuk.

Canada's official national sports are hockey in the winter and lacrosse in the summer. Hockey is a national pastime and the most popular spectator sport in the country. It is also the sport most played by Canadians, with 1.65 million participants in 2004. Canada's six largest metropolitan areas—Toronto, Montreal, Vancouver, Ottawa, Calgary, and Edmonton—have franchises in the National Hockey League (NHL), and there are more Canadian players in the NHL than from all other countries combined. Other popular spectator sports include curling and football; the latter is played professionally in the Canadian Football League (CFL). Golf, baseball, skiing, soccer, volleyball, and basketball are widely played at youth and amateur levels, but professional leagues and franchises are not widespread.

Canada has hosted several high-profile international sporting events, including the 1976 Summer Olympics in Montreal, the 1988 Winter Olympics in Calgary, and the 2007 FIFA U-20 World Cup. Canada was the host nation for the 2010 Winter Olympics in Vancouver and Whistler, British Columbia.

Language

Canada's two official languages are English and French. Official bilingualism is defined in the Canadian Charter of Rights and Freedoms, the Official Languages Act, and Official Language Regulations; it is applied by the Commissioner of Official Languages. English and French have equal status in federal courts, Parliament, and in all federal institutions. Citizens have the right, where there is sufficient demand, to receive federal government services in either English or French, and official-language minorities are guaranteed their own schools in all provinces and territories.

English and French are the mother tongues of 59.7% and 23.2% of the population respectively, and the languages most spoken at home by 68.3% and 22.3% of the population respectively. 98.5% of Canadians speak English or French (67.5% speak English only, 13.3% speak French only, and 17.7% speak both). English and French Official Language Communities, defined by First Official Language Spoken, constitute 73.0% and 23.6% of the population respectively.

The Charter of the French Language makes French the official language in Quebec. Although more than 85% of French-speaking Canadians live in Quebec, there are substantial Francophone populations in Ontario, Alberta, and southern Manitoba; Ontario has the largest French-speaking population outside Quebec. New Brunswick, the only officially bilingual province, has a French-speaking Acadian minority constituting 33% of the population. There are also clusters of Acadians in southwestern Nova Scotia, on Cape Breton Island, and through central and western Prince Edward Island.

Other provinces have no official languages as such, but French is used as a language of instruction, in courts, and for other government services in addition to English. Manitoba, Ontario, and Quebec allow for both English and French to be spoken in the provincial legislatures, and laws are enacted in both languages. In Ontario, French has some legal status but is not fully co-official. There are 11 Aboriginal language groups, made up of more than 65 distinct dialects. Of these, only Cree, Inuktitut and Ojibway have a large enough population of fluent speakers to be considered viable to survive in the long term. Several aboriginal languages have official status in the Northwest Territories. Inuktitut is the majority language in Nunavut, and one of three official languages in the territory.

Over six million people in Canada list a non-official language as their mother tongue. Some of the most common non-official first languages include Chinese (mainly Cantonese; 1,012,065 first-language speakers), Italian (455,040), German (450,570), Punjabi (367,505) and Spanish (345,345).

6

Mexico

The United Mexican States, commonly known as Mexico, is a federal constitutional republic in North America. It is bordered on the north by the United States; on the south and west by the Pacific Ocean; on the southeast by Guatemala, Belize, and the Caribbean Sea; and on the east by the Gulf of Mexico. Covering almost 2 million square kilometres (over 760,000 sq mi), Mexico is the fifth-largest country in the Americas by total area and the 14th largest independent nation in the world. With an estimated population of 111 million, it is the 11th most populous country and the most populous Hispanophone country on Earth. Mexico is a federation comprising thirty-one states and a Federal District, the capital city.

In Pre-Columbian Mesoamerica many cultures matured into advanced civilizations such as the Olmec, the Toltec, the Teotihuacan, the Zapotec, the Maya and the Aztec before the first contact with Europeans. In 1521, Spain conquered and colonized the territory, which was administered as the viceroyalty of New Spain which would eventually become Mexico as the colony gained independence in 1821. The post-independence period was characterized by economic instability, territorial secession and civil war, including foreign intervention, two empires and two long domestic dictatorships. The latter led to the Mexican Revolution in 1910, which culminated with the promulgation of the 1917 Constitution and the emergence of the country's current political system. Elections held in July 2000 marked the first time that an opposition party won the presidency from the Institutional Revolutionary Party (Spanish: Partido Revolucionario Institucional, PRI).

As a regional power, and since 1994 the first Latin American member of the Organisation for Economic Co-operation and Development (OECD), Mexico is firmly established as an upper middle-income country, and is considered a newly industrialized country and an emerging power. It has the 13th largest nominal GDP and the 11th largest by purchasing power parity. The economy is strongly linked to those of its North American Free Trade Agreement (NAFTA) partners, especially the United States, as well as tourism, being the world's tenth most visited country with over 21.4 million international arrivals. Mexico boasts a long tradition in the arts, renowned cuisine, and culture, as it ranks fifth in the world and first in the Americas with 29 UNESCO World Heritage Sites.

Etymology

After New Spain won independence from Spain, it was decided that the new country would be named after its capital, Mexico City, which was founded in 1524 on top of the ancient Aztec capital of Mexico-Tenochtitlan. The name comes from the Nahuatl language, but its meaning is not known. It has been suggested that it is derived from Mextli or Mcxihtli, a secret name for the god of war and patron of the Aztecs, Huitzilopochtli, in which case Mcxihco means "Place where Mcxihtli lives".

Another hypothesis suggests that the word Mcxihco derives from the mctztli ("moon"), xictli ("navel", "centre" or "son"), and the suffix-co (place), in which case it means "Place at the centre of the moon" or "Place at the centre of the Lake Moon", in reference to Lake Texcoco. The system of interconnected lakes, of which Texcoco was at the centre, had the form of a rabbit, the same image that the Aztecs saw in the moon. Tenochtitlan was located at the centre (or navel) of the lake (or rabbit/moon). Still another hypothesis suggests that it is derived from Mcctli, the goddess of maguey.

The name of the city was transliterated to Spanish as Mexico with the phonetic value of the x in Medieval Spanish, which represented the voiceless postalveolar fricative. This sound, as well as the voiced postalveolar fricative, represented by a j, evolved into a voiceless velar fricative during the sixteenth century. This led to the use of the variant Mejico in many publications in Spanish, most notably in Spain, whereas in Mexico and most other Spanish–speaking countries Mexico was the preferred spelling. In recent years the Real Academia Espanola, which regulates the Spanish language, determined that both variants are acceptable in Spanish but that the normative recommended spelling is Mexico. The majority of publications in all Spanish-speaking countries now adhere to the new norm, even though the alternative variant is still occasionally used. In English, the x in Mexico represents neither the original nor the current sound, but the consonant cluster.

The official name of the country has changed as the form of government has changed. On two occasions (1821–1823 and 1863–1867), the country was known as Imperio Mexicano (Mexican Empire). All three federal constitutions (1824, 1857 and 1917, the current constitution) used the name Estados Unidos Mexicanos—or the variants Estados Unidos mexicanos and Estados-Unidos Mexicanos, all of which have been translated as "United Mexican States". The term Republica Mexicana, "Mexican Republic" was used in the 1836 Constitutional Laws.

History

Campfire remains in the Valley of Mexico have been radiocarbon-dated to 21,000 BCE, and a few chips of stone tools have been found near the hearths, indicating the presence of humans at that time. Around 9,000 years ago, ancient indigenous peoples domesticated corn and initiated an agricultural revolution, leading to the formation of many complex civilizations. Between 1,800 and 300 BCE, many matured into advanced pre-Columbian Mesoamerican civilizations such as: the Olmec, the Teotihuacan, the Maya, the Zapotec, the Mixtec, the Toltec and the Aztec, which flourished for nearly 4,000 years before the first contact with Europeans.

These civilizations are credited with many inventions and advancements in fields such as architecture (pyramid-temples), mathematics, astronomy, medicine and theology. The Aztecs were noted for practicing human sacrifice on a large scale. At its peak, Teotihuacan, containing some of the largest pyramidal structures built in the pre-Columbian Americas, had a population of more than 150,000 people. Estimates of the population before the Spanish conquest range from 6 million to 25 million.

In the early 16th century, from the landing of Hernan Cortés, the Aztec civilization was invaded and conquered by the Spaniards. Unintentionally introduced by Spanish conquerors, smallpox ravaged Mesoamerica in the 1520s, killing millions of Aztecs, including the emperor, and was credited with the victory of Hernan Cortés over the Aztec empire. The territory became part of the Spanish Empire under the name of New Spain. Much of the identity, traditions and architecture of Mexico were created during the colonial period.

On September 16, 1810, independence from Spain was declared by priest Miguel Hidalgo y Costilla, in the small town of Dolores, Guanajuato. The first insurgent group was formed by Hidalgo, the Spanish viceregal army captain Ignacio Allende, the militia captain Juan Aldama and "La Corregidora" Josefa Ortiz de Dominguez. Hidalgo and some of his soldiers were captured and executed by firing squad in Chihuahua, on July 31, 1811. Following his death, the leadership was assumed by priest Jose Maria Morelos, who occupied key southern cities.

In 1813, the Congress of Chilpancingo was convened and, on November 6, signed the "Solemn Act of the Declaration of Independence of Northern America". Morelos was captured and executed on December 22, 1815. In subsequent years, the insurgency was near collapse, but in 1820 Viceroy Juan Ruiz de Apodaca sent an army under the criollo general Agustin de Iturbide against the troops of Vicente Guerrero. Instead, Iturbide approached Guerrero to join forces, and in 1821 representatives of the Spanish Crown and Iturbide signed the "Treaty of Cordoba", which recognized the independence of Mexico under the terms of the "Plan of Iguala".

Agustin de Iturbide immediately proclaimed himself emperor of the First Mexican Empire. A revolt against him in 1823 established the United Mexican States. In 1824, a Republican Constitution was drafted and Guadalupe Victoria became the first president of the newly born country. The first decades of the post-independence period were marked by economic instability, which led to the Pastry War in 1836, and a constant strife between liberales, supporters of a federal form of government, and conservadores, proposals of a hierarchical form of government.

General Antonio Lopez de Santa Anna, a centralist and two-time dictator, approved the Siete Leyes in 1836, a radical amendment that institutionalized the centralized form of government. When he suspended the 1824 Constitution, civil war spread across the country, and three new governments declared independence: the Republic of Texas, the Republic of the Rio Grande and the Republic of Yucatan.

Texas successfully achieved independence and was annexed by the United States. A border dispute led to the Mexican–American War, which began in 1846 and lasted for two years; the War was settled via the "Treaty of Guadalupe-Hidalgo", which forced Mexico to give up nearly half of its land to the U.S., including California and New Mexico. A much smaller transfer of territory in parts of southern Arizona and New Mexico — the Gadsden Purchase — occurred in 1854. The Caste War of Yucatan, the Mayan uprising that began in 1847, was one of the most successful modern Native American revolts. Maya rebels, or Cruzob, maintained relatively independent enclaves until the 1930s.

Dissatisfaction with Santa Anna's return to power led to the liberal "Plan of Ayutla", initiating an era known as La Reforma, after which a new Constitution was drafted in 1857 that established a secular state, federalism as the form of government, and several freedoms. As the conservadores refused to recognized it, the War of Reform began in 1858, during which both groups had their own governments. The war ended in 1861 with victory by the Liberals, led by Amerindian President Benito Juarez. In the 1860s Mexico underwent a military occupation by France, which established the Second Mexican Empire under the rule of Habsburg Archduke Ferdinand Maximilian of Austria with support from the Roman Catholic clergy and the conservadores, who later switched sides and joined the liberales. Maximilian surrendered, was tried on June 14 and was executed on June 19, 1867.

Porfirio Diaz, a republican general during the French intervention, ruled Mexico from 1876–1880 and then from 1884–1911 in five consecutive re-elections, period known as the

Porfiriato, characterized by remarkable economic achievements, investments in arts and sciences, but also of economic inequality and political repression.

20th Century

A likely electoral fraud that led to Diaz's fifth re-election sparked the 1910 Mexican Revolution, initially led by Francisco I. Madero.

Diaz resigned in 1911 and Madero was elected president but overthrown and murdered in a coup d'etat two years later directed by conservative general Victoriano Huerta. Event that re-ignited the civil war, involving figures such as Francisco Villa and Emiliano Zapata, who formed their own forces. A third force, the constitutional army led by Venustiano Carranza, managed to bring an end to the war, and radically amended the 1857 Constitution to include many of the social premises and demands of the revolutionaries into what was eventually called the 1917 Constitution. It is estimated that the war killed 900,000 of the 1910 population of 15 million.

Assassinated in 1920, Carranza was succeeded by another revolutionary hero, Alvaro Obregon, who in turn was succeeded by Plutarco Elias Calles. Obregon was re-elected in 1928 but assassinated before he could assume power. In 1929, Calles founded the National Revolutionary Party (PNR), later renamed the Institutional Revolutionary Party (PRI), and started a period known as the Maximato, which ended with the election of Lazaro Cardenas, who implemented many economic and social reforms, and most significantly expropriated the oil industry into PEMEX on March 18, 1938, but sparked a diplomatic crisis with the countries whose citizens had lost businesses by Cardenas radical measure.

Between 1940 and 1980, Mexico experienced a substantial economic growth that some historians call the "Mexican Miracle". Although the economy continued to flourish, social inequality remained a factor of discontent. Moreover, the PRI rule became increasingly authoritarian and at times oppressive (i.e.: the 1968 Tlatelolco massacre, which claimed the life of around 30–800 protesters).

Electoral reforms and high oil prices followed the administration of Luis Echeverria, mismanagement of these revenues led to inflation and exacerbated the 1982 Crisis. That year, oil prices plunged, interest rates soared, and the government defaulted on its debt. President Miguel de la Madrid resorted to currency devaluations which in turn sparked inflation.

In the 1980s, first cracks in the political monopolistic position of PRI were seen such as the election of Ernesto Ruffo Appel in Baja California and the 1988 electoral fraud, which prevented leftist candidate Cuauhtemoc Cardenas from winning the national presidential elections, who lost to Carlos Salinas de Gortari, leading to massive protests in Mexico City.

Salinas embarked on a program of neoliberal reforms which fixed the exchange rate, controlled inflation and culminated with the signing of the North American Free Trade Agreement (NAFTA), which came into effect on January 1, 1994. The same day, the Zapatista Army of National Liberation (EZLN) started a two-week-lived armed rebellion against the federal government, and has continued as a non-violent opposition movement against neoliberalism and globalization.

In December 1994, a month after Salinas was succeeded by Ernesto Zedillo, the Mexican economy collapsed, with a rapid rescue packaged authorized by U.S. President Bill Clinton and major macroeconomic reforms started by president Zedillo, the economy rapidly recovered and growth peaked at almost 7% by the end of 1999.

In 2000, after 71 years, the PRI lost a presidential election to Vicente Fox of the opposition National Action Party (PAN). In the subsequent presidential elections, Felipe Calderon from the PAN was declared the winner, with a razor-thin margin over leftist politician Andres Manuel Lopez Obrador of the Party of the Democratic Revolution (PRD). López Obrador, however, contested the election and pledged to create an "alternative government".

Administrative Divisions

The United Mexican States are a federation of thirty-one free and sovereign states, which form a union that exercises jurisdiction over the Federal District and other territories. Each state has its own Constitution and Congress, as well as a judiciary, and its citizens elect by direct voting, a governor for a six-year term, as well as representatives to their respective unicameral state congresses, for three-year terms.

The states are also divided into municipalities, the smallest administrative political entity in the country, governed by a mayor or municipal president (President municipal), elected by its residents by plurality. Municipalities can be further subdivided into non-autonomous boroughs or in semi-autonomous auxiliary presidencies.

Constitutionally, Mexico City, as the capital and seat of the federal powers, is the Federal District, a special political division that belongs to the federation as a whole and not to a particular state, and as such, has more limited local rule than the nation's states. Since 1987, it has progressively gained a greater degree of autonomy, and residents now elect a head of government and representatives of a Legislative Assembly directly. Unlike the states, the Federal District does not have a Constitution but a Statute of Government. Mexico City is coterminous and coextensive with the Federal District.

Geography and Climate

The geography of Mexico entails the physical and human geography of Mexico, a country situated in the Americas. Mexico is located at about 23° N and 102° W in the southern portion of North America. It is also located in a region known as Middle America. Mexico is bounded to the north by the United States (specifically, from west to east, by California, Arizona, New Mexico, and Texas), to the west and south by the Pacific Ocean, to the east by the Gulf of Mexico, and to the southeast by Belize, Guatemala, and the Caribbean Sea. The northernmost constituent of Latin America, it is the most populous Spanish-speaking country in the world.

Almost all of Mexico is on the North American Plate, with small parts of the Baja California Peninsula in the northwest on the Pacific and Cocos Plates. Some geographers include the portion east of the Isthmus of Tehuantepec including the Yucatan Peninsula within North America. This portion includes the five states of Campeche, Chiapas, Tabasco, Quintana Roo, and Yucatan, representing 12.1% of the country's total area. Alternatively, the Trans-Mexican Volcanic Belt may be said to delimit the region physiographically on the north. Geopolitically, Mexico is generally not considered part of Central America.

As well as numerous neighbouring islands, Mexican territory includes the more remote Isla Guadalupe and the Islas Revillagigedo in the Pacific Ocean. Mexico's total area covers 1,972,550 square kilometers, including approximately 6,000 square kilometers of islands in the Pacific Ocean, Gulf of Mexico, Caribbean Sea, and Gulf of California. On its north, Mexico shares a 5000-kilometer border with the United States. The meandering Rio Bravo del Norte (known as the Rio Grande in the United States) defines the border from Ciudad Juarez east to the Gulf of Mexico. A series of natural and artificial markers delineate the United States-Mexican border west from Ciudad Juarez to the Pacific Ocean. On its south, Mexico shares an 871 kilometre border with Guatemala and a 251-kilometer border with Belize.

Mexico has a 9,330 kilometre coastline, of which 7,338 kilometers face the Pacific Ocean and the Gulf of California, and the remaining 2,805 kilometers front the Gulf of Mexico and the Caribbean Sea. Mexico's exclusive economic zone (EEZ), which extends 200 nautical miles (370 km) off each coast, covers approximately 2.7 million square kilometers. The landmass of Mexico dramatically narrows as it moves in a southeasterly direction from the United States border and then abruptly curves northward before ending in the 500-kilometer-long Yucatan Peninsula. Indeed, the state capital of Yucatan, Merida, is farther north than Mexico City or Guadalajara.

Physical Features

Beginning approximately 50 kilometers from the United States border, the Sierra Madre Occidental extends 5000 kilometers south to the Rio Santiago, where it merges with the Cordillera Neovolcanica range that runs east-west across central Mexico. The Sierra Madre Occidental lies approximately 300 kilometers inland from the west coast of Mexico at its northern end but approaches to within fifty kilometers of the coast near the Cordillera Neovolcanica. The northwest coastal plain is the name given the lowland area between the Sierra Madre Occidental and the Gulf of California. The Sierra Madre Occidental averages 2,250 meters in elevation, with peaks reaching 3,000 meters. The Sierra Madre Oriental starts at the Big Bend region of the border with the U.S. state of Texas and continues 1,350 kilometers until reaching Cofre de Perote, one of the major peaks of the Cordillera Neovolcanica. As is the case with the Sierra Madre Occidental, the Sierra Madre Oriental comes progressively closer to the coastline as it approaches its southern terminus, reaching to within 75 kilometers of the Gulf of Mexico. The northeast coastal plain extends from the eastern slope of the Sierra Madre Oriental to the Gulf of Mexico. The median elevation of the Sierra Madre Oriental is 2,200 meters, with some peaks at 3,000 meters.

The Mexican altiplano, stretching from the United States border to the Cordillera Neovolcanica, occupies the vast expanse of land between the eastern and western sierra madres. A low east-west range divides the altiplano into northern and southern sections. These two sections, previously called the Mesa del Norte and Mesa Central, are now regarded by geographers as sections of one altiplano. The northern altiplano averages 1,100 meters in elevation and continues south from the Rio Bravo del Norte through the states of Zacatecas and San Luis Potosi. Various narrow, isolated ridges cross the plateaus of the northern altiplano. Numerous depressions dot the region, the largest of which is the Bolson de Mapimi. The southern altiplano is higher than its northern counterpart, averaging 2,000 meters in elevation. The southern altiplano contains numerous valleys originally formed by ancient lakes. Several of Mexico's most prominent cities, including Mexico City and Guadalajara, are located in the valleys of the southern altiplano.

One other significant mountain range, the Peninsular Ranges, cuts across the landscape of the northern half of Mexico. A southern extension of the California coastal ranges that parallel California's coast, the Mexican portion of the Peninsular Ranges extends from the United States border to the southern tip of the Baja California Peninsula, a distance of 1,430 kilometers. Peaks in the California system range in altitude from 2,200 meters in the north to only 250 meters near La Paz in the south. Narrow lowlands are found on the Pacific Ocean and the Gulf of California sides of the mountains. The Cordillera Neovolcanica is a belt 900 kilometers long and 130 kilometers wide, extending from the Pacific Ocean to the Gulf of Mexico.

The Cordillera Neovolcanica begins at the Rio Grande de Santiago and continues south to Colima, where it turns east along the nineteenth parallel to the central portion of the state of Veracruz. The region is distinguished by considerable seismic activity and contains Mexico's highest volcanic peaks. This range contains three peaks exceeding 5,000 meters: Pico de Orizaba (Citlaltepetl)—the third highest mountain in North America—and Popocatepetl and Iztaccihuatl near Mexico City. The Cordillera Neovolcanica is regarded as the geological dividing line between North America and Central America.

Several important mountain ranges dominate the landscape of southern and southeastern Mexico. The Sierra Madre del Sur extends 1,200 kilometers along Mexico's southern coast from the southwestern part of the Cordillera Neovolcanica to the nearly flat isthmus of Tehuantepec. Mountains in this range average 2,000 meters in elevation. The range averages 100 kilometers wide, but widens to 150 kilometers in the state of Oaxaca. The narrow southwest coastal plain extends from the Sierra Madre del Sur to the Pacific Ocean. The Sierra Madre de Oaxaca begins at Pico de Orizaba and extends in a southeasterly direction for 300 kilometers until reaching the isthmus of Tehuantepec. Peaks in the Sierra Madre de Oaxaca average 2,500 meters in elevation, with some peaks exceeding 3,000 meters. South of the isthmus of Tehuantepec, the Sierra Madre de Chiapas runs 280 kilometers along the Pacific Coast from the Oaxaca-Chiapas border to Mexico's border with Guatemala. Although average elevation is only 1,500 meters, one peak—Volcan de Tacuma—exceeds 4,000 meters in elevation. Finally, the Meseta Central de Chiapas extends 250 kilometers through the central part of Chiapas to Guatemala. The average height of peaks of the Meseta Central de Chiapas is 2,000 meters. The Chiapas central valley separates the Meseta Central de Chiapas and the Sierra Madre de Chiapas.

Mexico has nearly 150 rivers, two-thirds of which empty into the Pacific Ocean and the remainder of which flow into the Gulf of Mexico or the Caribbean Sea. Despite this apparent abundance of water, water volume is unevenly distributed throughout the country. Indeed, five rivers—the Usumacinta, Grijalva, Papaloapan, Coatzacoalcos, and Panuco—account for 52 % of Mexico's average annual volume of surface water. All five rivers flow into the Gulf of Mexico; only the Rio Panuco is outside southeastern Mexico, which contains approximately 15 % of national territory and 12 % of the national population. In contrast, northern and central Mexico, with 47 % of the national area and almost 60 % of Mexico's population, have less than 10 % of the country's water resources.

Seismic Activity

Situated atop three of the large tectonic plates that constitute the Earth's surface, Mexico is one of the most seismologically active regions on earth. The motion of these plates causes earthquakes and volcanic activity.

Most of the Mexican landmass rests on the westward moving North American plate. The Pacific Ocean floor off southern Mexico, however, is being carried northeast by the underlying motion of the Cocos Plate. Ocean floor material is relatively dense; when it strikes the lighter granite of the Mexican landmass, the ocean floor is forced under the landmass, creating the deep Middle America Trench that lies off Mexico's southern coast. The westward moving land atop the North American plate is slowed and crumpled where it meets the Cocos plate, creating the mountain ranges of southern Mexico. The subduction of the Cocos plate accounts for the frequency of earthquakes near Mexico's southern coast. As the rocks constituting the ocean floor are forced down, they melt, and the molten material is forced up through weaknesses in the surface rock, creating the volcanoes in the Cordillera Neovolcanica across central Mexico.

Areas of Mexico's coastline on the Gulf of California, including the Baja California Peninsula, are riding northwestward on the Pacific plate. Rather than one plate subducting, the Pacific and North American plates grind past each other, creating a slip fault that is the southern extension of the San Andreas fault in California. Motion along this fault in the past pulled Baja California away from the coast, creating the Gulf of California. Continued motion along this fault is the source of earthquakes in western Mexico.

Mexico has a long history of destructive earthquakes and volcanic eruptions. In September 1985, an earthquake measuring 8.1 on the Richter scale and centered in the subduction zone off Acapulco killed more than 4,000 people in Mexico City, more than 300 kilometers away. Volcan de Colima, south of Guadalajara, erupted in 1994, and El Chichon, in southern Mexico, underwent a violent eruption in 1983. Paricutin in northwest Mexico began as puffs of smoke in a cornfield in 1943; a decade later the volcano was 424 meters high. Although dormant for decades, Popocatepetl and Iztaccihuatl ("smoking warrior" and "white lady," respectively, in Nahuatl) occasionally send out puffs of smoke clearly visible in Mexico City, a reminder to the capital's inhabitants that volcanic activity is near. Popocatepetl showed renewed activity in 1995 and 1996, forcing the evacuation of several nearby villages and causing concern by seismologists and government officials about the effect that a large-scale eruption might have on the heavily populated region nearby.

Climate

The Tropic of Cancer effectively divides the country into temperate and tropical zones. Land north of the twenty-fourth parallel experiences cooler temperatures during the winter months. South of the twenty-fourth parallel, temperatures are fairly constant year round and vary solely as a function of elevation.

Areas south of the twentieth-fourth parallel with elevations up to 1,000 meters (3,281 ft) (the southern parts of both coastal plains as well as the Yucatan Peninsula), have a yearly median temperature between 24 and 28 °C (75.2 and 82.4 °F). Temperatures here remain high throughout the year, with only a 5 °C (9 °F) difference between winter and summer median temperatures. Although low-lying areas north of the twentieth-fourth parallel are hot and humid during the summer, they generally have lower yearly temperature averages (from 20 to 24 °C/68 to 75.2 °F) because of more moderate conditions during the winter.

Between 1,000 and 2,000 meters (3,281 and 6,562 ft), one encounters yearly average temperatures between 16 and 20 °C (60.8 and 68 °F). Towns and cities at this elevation south of the twenty-fourth parallel have relatively constant, pleasant temperatures throughout the year, whereas more northerly locations experience sizeable seasonal variations. Above

2,000 meters (6,562 ft), temperatures drop as low as an average yearly range between 8 and 12 °C (46.4 and 53.6 °F) in the Cordillera Neovolcanica. At 2,300 meters (7,546 ft), Mexico City has a yearly median temperature of 15 °C (59 °F) with pleasant summers and mild winters. Average daily highs and lows for May, the warmest month, are 26 and 12 °C (78.8 and 53.6 °F), and average daily highs and lows for January, the coldest month, are 19 and 6 °C (66.2 and 42.8 °F).

Rainfall varies widely both by location and season. Arid or semiarid conditions are encountered in the Baja California Peninsula, the northwestern state of Sonora, the northern altiplano, and also significant portions of the southern altiplano. Rainfall in these regions averages between 300 and 600 millimeters (11.8 and 23.6 in) per year, although even less in some areas, particularly in Baja California Norte. Average rainfall totals are between 600 and 1,000 millimeters (23.6 and 39.4 in) in most of the major populated areas of the southern altiplano, including Mexico City and Guadalajara. Low-lying areas along the Gulf of Mexico receive in excess of 1,000 millimeters (39.4 in) of rainfall in an average year, with the wettest region being the southeastern state of Tabasco, which typically receives approximately 2,000 millimeters (78.7 in) of rainfall on an annual basis. Parts of the northern altiplano and high peaks in the Sierra Madre Occidental and the Sierra Madre Oriental occasionally receive significant snowfalls.

Mexico has pronounced wet and dry seasons. Most of the country experiences a rainy season from June to mid-October and significantly less rain during the remainder of the year. February and July generally are the driest and wettest months, respectively. Mexico City, for example, receives an average of only 5 millimeters (0.2 in) of rain during February but more than 160 millimeters (6.3 in) in July. Coastal areas, especially those along the Gulf of Mexico, experience the largest amounts of rain in September. Tabasco typically records more than 300 millimeters (11.8 in) of rain during that month. A small coastal area of northwestern coastal Mexico around Tijuana has a Mediterranean climate with considerable coastal fog and a rainy season that occurs in winter. Mexico lies squarely within the hurricane belt, and all regions of both coasts are susceptible to these storms from June through November. Hurricanes on the Pacific coast are less frequent and often less violent than those affecting Mexico's eastern coastline. Several hurricanes per year strike the Caribbean and Gulf of Mexico coastline, however, and these storms bring high winds, heavy rain, extensive damage, and occasional loss of life. Hurricane Gilbert passed directly over Cancun in September 1988, with winds in excess of 200 kilometers per hour (124 mph), producing major damage to hotels in the resort area. It then struck northeast Mexico, where flooding from the heavy rain killed dozens in the Monterrey area and caused extensive damage to livestock and vegetable crops.

Environmental Conditions

Mexico faces significant environmental challenges affecting almost every section of the country. Vast expanses of southern and southeastern tropical forests have been denuded for cattle-raising and agriculture. For example, tropical forests covered almost half of the state of Tabasco in 1940 but less than 10% by the late 1980s. During the same period, pastureland increased from 20 to 60% of the state's total area. Analysts reported similar conditions in other tropical sections of Mexico. Deforestation has contributed to serious levels of soil erosion nationwide. In 1985 the government classified almost 17% of all land as totally eroded, 31% in an accelerated state of erosion, and 38% demonstrating signs of incipient erosion. Mexico has developed a Biodiversity Action Plan to address issues of endangered species and habitats that merit protection.

Soil destruction is particularly pronounced in the north and northwest, with more than 60% of land considered in a total or accelerated state of erosion. Fragile because of its semiarid and arid character, the soil of the region has become increasingly damaged through excessive cattle-raising and irrigation with waters containing high levels of salinity. The result is a mounting problem of desertification throughout the region.

Mexico's vast coastline faces a different, but no less difficult, series of environmental problems. For example, inadequately regulated petroleum exploitation in the Coatzacoalcos-Minatitlan zone in the Gulf of Mexico has caused serious damage to the waters and fisheries of Rio Coatzacoalcos. The deadly explosion that racked a working-class neighbourhood in Guadalajara in April 1992 serves as an appropriate symbol of environmental damage in Mexico. More than 1,000 barrels of gasoline seeped from a corroded Mexican Petroleum (Petroleos Mexicanos—Pemex) pipeline into the municipal sewer system, where it combined with gases and industrial residuals to produce a massive explosion that killed 190 persons and injured nearly 1,500 others.

Mexico City confronts authorities with perhaps their most daunting environmental challenge. Geography and extreme population levels have combined to produce one of the world's most polluted urban areas. Mexico City sits in a valley surrounded on three sides by mountains, which serve to trap contaminants produced by the metropolitan area's 15 million residents. One government study in the late 1980s determined that nearly 5 million tons of contaminants were emitted annually in the atmosphere, a tenfold increase over the previous decade. Carbons and hydrocarbons from the region's more than 3 million vehicles account for approximately 80% of these contaminants, with another 15%, primarily of sulfur and nitrogen, coming from industrial plants. The resulting dangerous mix is responsible for a wide range of respiratory illnesses. One study of twelve urban areas worldwide in the mid-1980s concluded that the residents of Mexico City had the highest levels of lead and cadmium in their blood. The volume of pollutants from Mexico City has damaged the surrounding ecosystem as well. For example, wastewater from Mexico City that flows north and is used for irrigation in the state of Hidalgo has been linked to congenital birth defects and high levels of gastrointestinal diseases in that state.

Beginning in the mid-1980s, the government enacted numerous antipollution policies in Mexico City with varied degrees of success. Measures such as vehicle emissions inspections, the introduction of unleaded gasoline, and the installation of catalytic converters on new vehicles helped reduce pollution generated by trucks and buses. In contrast, one of the government's most prominent actions, the No Driving Day program, may have inadvertently contributed to higher pollution levels. Under the program, metropolitan area residents were prohibited from driving their vehicles one day each work week based on the last number of their license plate. However, those with the resources to do so purchased additional automobiles to use on the day their principal vehicle was prohibited from driving, thus adding to the region's vehicle stock. Thermal inversions reached such dangerous levels at various times in the mid-1990s that the government declared pollution emergencies, necessitating sharp temporary cutbacks in vehicle use and industrial production.

General Indicators

Climate: varies from tropical to desert.

Terrain: high, rugged mountains; low coastal plains; high plateaus; desert.

Elevation extremes:

- lowest point: Laguna Salada-10 m
- highest point: Pico de Orizaba volcano 5,610 m.

Natural resources: petroleum, silver, copper, gold, lead, zinc, natural gas and timber.

Land use:

- arable land: 12%
- permanent crops: 1%
- permanent pastures: 39%
- forests and woodland: 26%
- other: 22% (1993 est.)

Irrigated land: 61,000 km² (1993 est.)

Natural hazards: Tsunamis along the Pacific coast, volcanoes and destructive earthquakes in the centre and south, and hurricanes on the Gulf of Mexico and Caribbean coasts.

Environment – current issues: Natural fresh water resources scarce and polluted in north, inaccessible and poor quality in centre and extreme southeast; raw sewage and industrial effluents polluting rivers in urban areas; deforestation; widespread erosion; desertification; serious air pollution in the national capital and urban centers along the US-Mexico border.

Environment – international agreements: Party to: Biodiversity, Climate Change, Desertification, Endangered

Species, Hazardous Wastes, Law of the Sea, Marine Dumping, Marine Life Conservation, Nuclear Test Ban, Ozone Layer Protection, Ship Pollution, Wetlands, Whaling and Kyoto Protocol.

Biodiversity

Mexico is one of the 18 megadiverse countries of the world. With over 200,000 different species, Mexico is home of 10–12% of the world's biodiversity. Mexico ranks first in biodiversity in reptiles with 707 known species, second in mammals with 438 species, fourth in amphibians with 290 species, and fourth in flora, with 26,000 different species. Mexico is also considered the second country in the world in ecosystems and fourth in overall species. Approximately 2,500 species are protected by Mexican legislations.

The Mexican government created the National System of Information about Biodiversity, in order to study and promote the sustainable use of ecosystems. Deforestation is one of the most serious environmental issues in Mexico, with more than one million hectares of forest being lost each year. As of 2002, Mexico had the second fastest rate of deforestation in the world, second only to Brazil. The government has taken another initiative in the late 1990s to expand the people's knowledge, interest and use of the country's esteemed biodiversity, through the Comision National para el Conocimiento y Uso de la Biodiversidad.

In Mexico, 170,000 square kilometres (65,637 sq mi) are considered "Protected Natural Areas." These include 34 reserve biospheres (unaltered ecosystems), 64 national parks, 4 natural monuments (protected in perpetuity for their aesthetic, scientific or historical value), 26 areas of protected flora and fauna, 4 areas for natural resource protection (conservation of soil, hydrological basins and forests) and 17 sanctuaries (zones rich in diverse species).

The discovery of the Americas brought to the rest of the world many widely used food crops and edible plants. Some of Mexico's native culinary ingredients include:Chocolate, Avocado, Tomato, Maize, Vanilla, Guava, Chayote, Epazote, Camote, Jicama, Nopal, Nucchini, Tejocote, Huitlacoche, Sapote, Mamey Sapote, many varieties of beans, and an even greater variety of chiles, such as the Habanero and the Xalapeno. Most of these names come from indigenous languages like Nahuatl.

Government and Politics

The United Mexican States are a federation whose government is representative, democratic and republican based on a presidential system according to the 1917 Constitution. The constitution establishes three levels of government: the federal Union, the state governments and the municipal governments. All officials at the three levels are elected by voters through first-past-the-post plurality, proportional representation or are appointed by other elected officials.

The federal government is constituted by the Powers of the Union, the three separate branches of government:

Legislature

Legislative: the bicameral Congress of the Union, composed of a Senate and a Chamber of Deputies, which makes federal law, declares war, imposes taxes, approves the national budget and international treaties, and ratifies diplomatic appointments.

Executive

Executive: the President of the United Mexican States, who is the head of state and government, as well as the commander-in-chief of the Mexican military forces. The President also appoints the Cabinet and other officers. The President is responsible for executing and enforcing the law, and has the authority of vetoing bills.

Judiciary

Judiciary: The Supreme Court of Justice, comprised by eleven judges appointed by the President with Senate approval, who interpret laws and judge cases of federal competency. Other institutions of the judiciary are the Electoral Tribunal, collegiate, unitary and district tribunals, and the Council of the Federal Judiciary.

All elected executive officials are elected by plurality (first-past-the-post). Seats to federal and state legislatures are elected by a system of parallel voting that includes plurality and proportional representation. The Chamber of Deputies of the Congress of the Union is conformed by 300 deputies elected by plurality and 200 deputies by proportional representation with closed party lists for which the country is divided into 5 electoral constituencies or circumscriptions.

The Senate is conformed by a total of 128 senators: 64 senators, two for each state and two for the Federal District, elected by plurality in pairs; 32 senators assigned to the first minority or first-runner up (one for each state and one for the Federal District), and 32 are assigned by proportional representation with closed party lists for which the country conforms a single electoral constituency.

According to the constitution, all constituent states of the federation must have a republican form of government composed of three branches: the executive, represented by a governor and an appointed cabinet, the legislative branch constituted by a unicameral congress and the judiciary, which will include called state Supreme Court of Justice. They also have their own civil and judicial codes.

In the 2009-2012 Congress of the Union, seven parties are therein represented; four of them, however, have not received neither in this nor in previous congresses more than 4% of the national votes. The other three parties have historically been the dominant parties in Mexican politics:

- National Action Party (Partido Accion National, PAN): a centre-right conservative party founded in 1939. PAN has gained plurality, but not absolute majority in several parliamentary elections. In 2000, it gained the presidency for the first time. It belongs to the Christian Democrat Organization of America.
- Institutional Revolutionary Party (Partido Revolucionario Institucional, PRI): a centre-left party that ascribes to social democracy—it is a member of Socialist International— founded in 1929 to unite all the factions of the Mexican Revolution. Prominent left-wing Mexican politicians have been members of the party. Having dominated Mexican politics since the Revolution, PRI includes diverse factions including some centre-right members.
- Party of the Democratic Revolution (Partido de la Revolution Democratica, PRD): a left-wing party,

founded in 1989 as the successor of the coalition of socialists and liberal parties, the National Democratic Front that had presented the candidacy of Cuauhtemoc Cardenas in the controversial 1988 elections.

The PRI held an almost hegemonic power in Mexican politics since 1929. Since 1977 consecutive electoral reforms allowed opposition parties to win more posts at the local and federal level. This process culminated in the 2000 presidential elections in which Vicente Fox, candidate of the PAN, became the first non-PRI president to be elected in 71 years.

In 2006, Felipe Calderón of the PAN faced Andres Manuel López Obrador of the PRD in a very close election (0.58% difference), by simple plurality—the Mexican electoral system does not include runoff voting. López Obrador contested the elections, but on September 6, 2006, Felipe Calderón was declared President-elect by the Electoral Tribunal. His cabinet was sworn in at midnight on December 1, 2006 and Calderón was handed the presidential band by outgoing Vicente Fox at Los Pinos. He was officially sworn as President on the morning of December 1, 2006 in Congress.

Foreign Relations

The foreign policy of Mexico is directed by the President and managed through the Ministry of Foreign Affairs, whose constitutionally recognized principles are: respect for international law and legal equality of states, their sovereignty and independence, non-intervention, peaceful resolution of conflicts and promotion of collective security through active participation in international organizations.

Since the 1930s, the Estrada Doctrine has served as a crucial complement to these principles. The foreign relations of Mexico have been focused primarily on the United States and its historically tied neighbours in Latin America and the Caribbean. In the 20th century, Mexico developed a foreign policy based on hemispheric prestige. However, in the 2000s, former President Vicente Fox adopted a new foreign policy that called for an openness and an acceptance of criticism from the international community and the increase of Mexican involvement in foreign affairs, as well as a further integration towards its northern neighbours. A greater priority to Latin

America and the Caribbean has been given in the administration of President Felipe Calderón.

In addition, since the 1990s Mexico has sought a reform of the United Nations Security Council and its working methods with the support of Canada, Italy, Pakistan and other nine countries, which form a group informally called the Coffee Club. As an regional and emerging power, Mexico has a significant global presence and is a member of several international organizations and forums such as the United Nations, the Organization of American States, the G8+5, the G-20 major economies, the Asia-Pacific Economic Cooperation and the Organisation for Economic Co-operation and Development.

Military

Mexico has the third-largest defence budget in Latin America, with reported annual military expenditures of US $24.944 billion or about 1.6% of GDP. Since the 1990s, when the military escalated its role in the war on drugs, increasing importance has been placed on acquiring airborne surveillance platforms, aircraft, helicopters, digital war-fighting technologies, urban warfare equipment and rapid troop transport.

The Mexican Military has two branches: the Mexican Army (which includes the Mexican Air Force), and the Mexican Navy. The Mexican armed forces maintain significant infrastructure, including facilities for design, research, and testing of weapons, vehicles, aircraft, naval vessels, defence systems and electronics; military industry manufacturing centers for building such systems, and advanced naval dockyards that build heavy military vessels and advanced missile technologies.

These facilities have a significant employment and economic impact. In recent years, Mexico has improved its training techniques, military command and information structures and has taken steps to becoming more self-reliant in supplying its military by designing as well as manufacturing its own arms, missiles, aircraft, vehicles, heavy weaponry, electronics, defence systems, armour, heavy military industrial equipment and heavy naval vessels.

Historically, Mexico has remained neutral in international conflicts with the exception of World War II. However, in recent years some political parties have proposed an amendment of the Constitution in order to allow the Mexican army, air force or navy to collaborate with the United Nations in peacekeeping missions, or to provide military help to countries that officially ask for it.

Law Enforcement

Public security is enacted at the three levels of government, each of which has different prerogatives and responsibilities. Local and state police department are primarily in charge of law enforcement, whereas the Federal Preventive Police is in charge of specialized duties. All levels report to the Secretaría de Seguridad Pública (Secretariat of Public Security). The General Attorney's Office (Procuraduria General de la República, PGR) is the executive power's agency in charge of investigating and prosecuting crimes at the federal level, mainly those related to drug and arms trafficking, espionage, and bank robberies. The PGR operates the Federal Investigations Agency (Agencia Federal de Investigacion, AFI) an investigative and preventive agency.

While the government respects the human rights of most citizens, serious abuses of power have been reported in security operations in indigenous communities and poor urban neighbourhoods. The National Human Rights Commission has had little impact in reversing this trend, engaging mostly in documentation but failing to use its powers to issue public condemnations to the officials who ignore its recommendations. By law, all defendants have the rights that assure them fair trials and human treatment; however, the system is overburdened and overwhelmed with several problems.

Despite the efforts of the authorities to fight crime and fraud, few Mexicans have strong confidence in the police or the judicial system, and therefore, few crimes are actually reported by the citizens. In 2008, president Calderón proposed a major reform of the judicial system, which was approved by the Congress of the Union, which included oral trials, the presumption of innocence for defendants, the authority of local police to investigate crime—until then a prerogative of special police units—and several other changes intended to speed up trials.

Crime

Total crimes per capita average 12 per 1,000 people in Mexico, ranking 39 in a survey of 60 countries. Violent crime is a critical issue in Mexico; with a rate of homicide varying from 11 to 14 per 100,000 inhabitants. Drug-traffic and narco-related activities are a major concern in Mexico. The Mexican drug cartels have as many as 100,000 foot soldiers, which is about the size of the Mexican army. Drug cartels are active in the shared border with the US and police corruption and collusion with drug cartels is a crucial problem.

Current president Felipe Calderón made abating drug-trafficking one of the top priorities of his administration. In a very controversial move, Calderón deployed military personnel to cities where drug cartels operate. While this move has been criticized by the opposition parties and the National Human Rights Commission, its effects have been praised by the Bureau for International Narcotics and Law Enforcement Affairs as having obtained "unprecedented results..." with "many important successes". In October 2007, the president Calderón and US president George W. Bush announced the Merida Initiative a historic plan of law enforcement cooperation between the two countries.

Economy

The economy of Mexico is the 13th largest in the world in nominal terms; and the 11th by purchasing power parity, according to the World Bank. Since the 1994 crisis, administrations have improved the country's macroeconomic fundamentals. Mexico was not significantly influenced by the recent 2002 South American crisis, and maintained positive, although low, rates of growth after a brief period of stagnation in 2001. However, Mexico was one of the Latin American nations most affected by the 2008 recession with its Gross Domestic Product contracting by more than 6%. Moody's (in March 2000) and Fitch IBCA (in January 2002) issued investment-grade ratings for Mexico's sovereign debt. In spite of its unprecedented macroeconomic stability, which has reduced inflation and interest rates to record lows and has increased per capita income, enormous gaps remain between the urban and the rural population, the northern and southern states, and the rich and the poor. Some of the government's challenges include the upgrade of infrastructure, the modernization of the tax system and labour laws, and the reduction of income inequality.

The economy contains rapidly developing modern industrial and service sectors, with increasing private ownership. Recent administrations have expanded competition in ports, railroads, telecommunications, electricity generation, natural gas distribution and airports, with the aim of upgrading infrastructure. As an export-oriented economy, more than 90% of Mexican trade is under free trade agreements (FTAs) with more than 40 countries, including the European Union, Japan, Israel, and much of Central and South America. The most influential FTA is the North American Free Trade Agreement (NAFTA), which came into effect in 1994, and was signed in 1992 by the governments of the United States, Canada and Mexico. In 2006, trade with Mexico's two northern partners accounted for almost 90% of its exports and 55% of its imports. Recently, the Congress of the Union approved important tax, pension and judicial reforms, and reform to the oil industry is currently being debated. According to the Forbes Global 2000 list of the world's largest companies in 2008, Mexico had 16 companies in the list.

History

Following five decades of political turbulence following the independence of Mexico, the four consecutive administrations of president Porfirio Diaz (during the last quarter of the nineteenth century) brought unprecedented economic growth. This growth was accompanied by foreign investment and European immigration, the development of an efficient railroad network and the exploitation of the country's natural resources. Annual economic growth between 1876 and 1910 averaged 3.3%. Political repression and fraud, as well as huge income inequalities exacerbated by the land distribution system based on latifundios, in which large haciendas were owned by a few but worked by millions of underpaid peasants living in precarious conditions, led to the Mexican Revolution (1910–1917), an armed conflict that drastically transformed Mexico's political, social, cultural, and economical structure during the twentieth century under a premise of social democracy. The war itself, however, left a harsh toll in the economy and population, which decreased over the 11-year period between 1910 and 1921. The reconstruction of the country was to take place in the following decades.

The period from 1930 to 1970 was dubbed by economic historians as the "Mexican Miracle", a period of economic growth spurred by a model of import substitution industrialization (ISI) which protected and promoted the development of national industries. Through the ISI model, the country experienced an economic boom through which industries rapidly expanded their production. Important changes in the economic structure included free land distribution to peasants under the concept of ejido, the nationalization of the oil and railroad companies, the introduction of social rights into the constitution, the birth of large and influential labour unions, and the upgrading of infrastructure. While population doubled from 1940 to 1970, GDP increased sixfold.

The ISI model had reached its peaked in the late 1960s. During the 1970s, the administrations of Echeverria and López Portillo, tried to include social development in their policies, an effort that entailed more public spending. With the discovery of vast oil fields in a time in which oil prices were surging and international interest rates were low-and even negative-the government decided to borrow from international capital markets to invest in the state-owned oil company, which in turn seemed to provide a long-run income source to promote social welfare. In fact, this method produced a remarkable growth in public expenditure, and president López Portillo announced that the time had come to learn to "manage prosperity" as Mexico multiplied its oil production to become the world's fourth largest exporter.

In the period of 1981–1982 the international panorama changed abruptly: oil prices plunged and interest rates rose. In 1982, president López Portillo, just before ending his administration, suspended payments of foreign debt, devalued the peso and nationalized the banking system, along with many other industries that were severely affected by the crisis, among them the steel industry. While import substitution had produced an era of industrialization in previous decades, by the 1980s it was evident that protracted protection had produced an uncompetitive industrial sector with low productivity gains. President de la Madrid was the first of a series of presidents that began to implement neoliberal reforms. After the crisis of 1982, lenders were unwilling to return to Mexico and, in order to keep the current account in balance, the government resorted to currency devaluations, which in turn sparked unprecedented inflation, which reached a historic high in 1987 at 159.7%.

The first step toward the liberalization of trade was Mexico's signature of the General Agreement on Tariffs and Trade (GATT) in 1986. During the Salinas administration many state-owned companies were privatized. In 1992, the North American Free Trade Agreement was signed between the United States, Canada and Mexico, and after the signature of two additional supplements on environments and labour standards, it came into effect on January 1, 1994. Salinas also introduced strict price controls and negotiated smaller minimum wage increments with labour unions with the aim of curbing inflation. While his strategy was successful in reducing inflation, growth averaged only 2.8 percent a year. Moreover, by fixing the exchange rate, the peso became rapidly overvalued while consumer spending increased, causing the current account deficit to reach 7% of GDP in 1994. The deficit was financed through tesobonos a type of public debt instrument that reassured payment in dollars. The Chiapas uprising, and the assassinations of the ruling party's presidential candidate, Luis Donaldo Colosio and the Secretary-General of the party and brother of the Assistant-Attorney General Jose Francisco Ruiz Massieu in 1994, sent a disquieting message to investors.

Public debt holders rapidly sold their tesobonos, depleting the Central Bank's reserves, while portfolio investments, which had made up 90% of total investment flows, left the country as fast as they had come in. This unsustainable situation eventually forced the entrant Zedillo administration to abandon the fixed exchange rate. The peso sharply devalued and the country entered into an economic crisis in December 1994. The boom in exports, as well as an international rescue package crafted by American president Bill Clinton, helped cushion the crisis. In less than 18 months, the economy was growing again, and annual rate growth averaged 5.1 percent between 1995 and 2000.

President Zedillo and president Fox continued with trade liberalization and during his administrations several FTAs were signed with Latin American and European countries, Japan and Israel, and both strove to maintain macroeconomic stability. Thus, Mexico became one of the most open countries in the world to trade, and the economy base shifted accordingly. Total trade with the United States and Canada tripled, and total exports and imports almost quadrupled between 1991 and 2003. The nature of foreign investment also changed from portfolio to foreign-direct investment (FDI).

Macroeconomic, Financial and Welfare Indicators

Main Indicators

Mexico's Gross Domestic Product (GDP) in purchasing power parity (PPP) was estimated at US $1.353 trillion in 2006, and $886.4 billion in nominal exchange rates. As such, its standard of living, as measured in GDP in PPP per capita was US $12,500. The World Bank reported in 2007 that the country's Gross National Income in market exchange rates was the second highest in Latin America, after Brazil at US $820.319 billion, which lead to the highest income per capita in the region at $7,830. As such, Mexico is now firmly established as an upper middle-income country. After the slowdown of 2001 the country has recovered and has grown 4.2, 3.0 and 4.8 percent in 2004, 2005 and 2006, even though it is considered to be well below Mexico's potential growth.

The Mexican currency is the peso. One peso is divided into 100 centavos (cents). MXN replaced MXP in 1993 at a rate of 1000 MXP per 1 MXN. The exchanged rate has remained stable since 1998, oscillating between 9.20 and 11.50 MXN per USD. Interest rates in 2007 were situated at around 7 percent, having reached a historic low in 2002 below 5 percent. Inflation rates are also at historic lows; the inflation rate in Mexico in 2006 was 4.1 percent, and 3 percent by the end of 2007. Unemployment rates are the lowest of all OECD member countries at 3.2 percent. However, underemployment is estimated at 25 percent. Mexico's Human development index was reported at 0.829, (comprising a life expectancy index of 0.84, an education index of 0.86 and a GDP index of 0.77), ranking 52 in the world within the group of high-development.

Poverty

After the 1994–1995 economic crisis, probably the most severe in the country's history, 50% of the population fell into poverty. A rapid growth in exports propitiated by NAFTA and other trade agreements, and the restructuring of the macroeconomic finances initiated during Zedillo's and continued during Fox's administration had significant results in the reduction of the poverty rate: according to the World Bank, poverty was reduced from 24.2% in 2000 to 17.6% in 2004. Most of this reduction was achieved in rural communities whose rate of poverty declined from 42% to 27.9% in the 2000–2004 period, although urban poverty stagnated at 12%. According to the World Bank, in 2004, 17.6% of Mexico's population lived in extreme poverty, while 21% lived in moderated poverty. The CIA Factbook, on the other hand, reported that 13.8% of the population was under the poverty line, as measured in food-based poverty.

Remittances

Remittances, or contributions sent by Mexicans living abroad, mostly in the United States, to their families at home in Mexico, are a substantial and growing part of the Mexican economy; they comprised $18 billion in 2005. In 2004, they became the second largest source of foreign income after crude oil exports, roughly equivalent to foreign direct investment (FDI) and larger than tourism expenditures; and represented 2.5 percent of the nation's Gross Domestic Product. The growth of remittances has been remarkable: they have more than doubled since 1997. Recorded remittance transactions exceeded 41 million in 2003, of which 86 percent were made by electronic transfer. The Mexican government, cognizant of the economic viability of immigrant workers, began issuing an upgraded version of the Matricula Consular de Alta Seguridad (MACS, High Security Consular Identification), an identity document issued at Mexican consulates abroad. This document is now accepted as a valid identity card in 32 US states, as well as thousands of police agencies, hundreds of cities and counties, as well as banking institutions.

The main receptors of remittances in 2004 were the states of Michoacan, Guanajuato, Jalisco, Mexico and Puebla, which jointly captured 45% of total remittances in that year. Several state governments, with the support of the federal government, have implemented programs to use part of the remittances to finance public works. This program, called Dos por Uno (Two for every one) is designed in a way that for each peso contributed by migrants from their remittances, the state and the federal governments will invest two pesos in building infrastructure at their home communities.

Regional Economies

Regional disparities and income inequality continue to be a problem in Mexico. While all constituent states of the federation have a Human Development Index (HDI) superior to 0.70 (medium to high development), northern and central states have higher levels of HDI than the southern states. Nuevo Leon, Campeche and the Federal District have HDI levels similar to European countries, whereas that of Oaxaca and Chiapas is similar to that of Syria or Egypt. At the municipal level, disparities are even greater: San Pedro Garza Garcia in Nuevo Leon has an HDI similar to that of Spain, whereas, Metlatonoc in Guerrero, would have an HDI similar to that of Malawi. The majority of the federal entities with high development (superior to 0.80) are located in the northern region (with the exception of Colima, Jalisco, Aguascalientes, the Federal District, Queretaro, as well as the southeastern states of Quintana Roo and Campeche). The less developed states (with medium development in terms of HDI, superior to 0.70) are located at the southern Pacific coast (with the exception of Veracruz).

In terms of share in GDP per sector (in 2004), the largest contributors in agriculture are Jalisco (9.7%), Sinaloa (7.7%) and Veracruz (7.6%); the greatest contributors in industrial production are the Federal District (15.8%), State of Mexico (11.8%) and Nuevo Leon (7.9%); the greatest contributors in the service sector are also the Federal District (25.3%), State of Mexico (8.9%) and Nuevo Leon (7.5%).

Since the 1980s, the economy has slowly become less centralized; the annual rate of GDP growth of the Federal District from 2003–2004 was the smallest of all federal entities at a mere 0.23%, with drastic drops in the agriculture and industrial sectors. Nonetheless, it still accounts for 21.8% of the nation's GDP. The states with the highest GDP growth rates are Quintana Roo (9.04%), Baja California (8.89%), and San Luis Potosi (8.18%). In 2000, the federal entities with the highest GDP per capita in Mexico were the Federal District (US $17,696), Campeche (US $13,153) and Nuevo Leon (US $13,033); the states with the lowest GDP per capita were Chiapas (US $3,302), Oaxaca (US $3,489) and Guerrero (US $4,112).

Components of the Economy

Gross Domestic Product (GDP) in purchasing power parity (PPP) in 2006 was estimated at US $1.134 trillion, and GDP per capita in PPP at US $10,600. The service sector is the largest component of GDP at 70.5%, followed by the industrial sector at 25.7% (2006 est.). Agriculture represents only 3.9% of GDP (2006 est.). Mexican labour force is estimated at 38 million of which 18% is occupied in agriculture, 24% in the industry sector and 58% in the service sector (2003 est.).

Agriculture and Food Production

Food and Agriculture

Product	*Quantity (Tm)*	*World Rank*
Avocados	1,040,390	1
Onions and chayote	1,130,660	1
Limes and lemons	1,824,890	1
Sunflower seed	212,765	1
Dry fruits	95,150	2
Papaya	955,694	2
Chillies and peppers	1,853,610	2
Whole beans	93 000	3
Oranges	3,969,810	3
Anise, badian, fennel	32 500	3
Chicken meat	2,245,000	3
Asparagus	67,247	4
Mangoes	1.503.010	4
Corn	20,000,000	4

Source:FAO

After the Mexican Revolution Mexico began an agrarian reform, based on the 27th article of the Mexican Constitution than included transfer of land and/or free land distribution to peasants and small farmers under the concept of the ejido. This program was further extended during president Cardenas administration during the 1930s and continued into the 1960s at varying rates.

The cooperative agrarian reform, which guaranteed small farmers a means of subsistence livelihood, also caused land fragmentation and lack of capital investment, since commonly held land could not be used as collateral. In an effort to raise rural productivity and living standards, this constitutional article was amended in 1992 to allow for the transfer of property rights of the communal lands to farmers cultivating it. With the ability to rent or sell it, a way was open for the creation of larger farms and the advantages of economies of scale. Large mechanized farms are now operating in some northeastern states (mainly in Sinaloa). However, privatization of ejidos continues to be very slow in the central and southern states where the great majority of peasants produce only for subsistence.

Up until the 1990s, the government encouraged the production of basic crops (mainly corn and beans) by maintaining support prices and controlling imports through the National Company for Popular Subsistence (CONASUPO). With trade liberalization, however, CONASUPO was to be gradually dismantled and two new mechanisms were implemented: Alianza and Procampo. Alianza provides income payments and incentives for mechanization and advanced irrigation systems. Procampo is an income transfer subsidy to farmers. This support program provides 3.5 million farmers who produce basic commodities (mostly corn), and which represent 64% of all farmers, with a fixed income transfer payment per unit of area of cropland. This subsidy increased substantially during president Fox's administration, mainly to white corn producers in order to reduce the amount of imports from the United States. This program has been successful, and in 2004, roughly only 15% of corn imports are white corn –the one used for human consumption and the type that is mostly grown in Mexico– as opposed to 85% of yellow and crashed corn –the one use for feeding livestock, and which is barely produced in Mexico.

Importance of Agriculture to Mexico's Economy

Agriculture, as a percentage of GDP, has been steadily declining, and now resembles that of developed nations, in that it plays a smaller role in the economy. In 2006, agriculture accounted for only 3.9% of GDP, down from 7% in 1980, and 25% in 1970. Nonetheless, given the historic structure of ejidos, it still employs a considerably high percentage of the work force: 18% in 2003, mostly of which grows basic crops for subsistence, compared to 2–5% in developed nations in which production is highly mechanized.

Crops

In spite of being a staple in the Mexican diet, Mexico's comparative advantage in agriculture is not in corn, but in horticulture, tropical fruits, and vegetables. Negotiators of NAFTA expected that through liberalization and mechanization of agriculture two-thirds of Mexican corn producers would naturally shift from corn production to horticultural and other labour-intensive crops such as fruits, nuts, vegetables, coffee and sugar cane.

While horticultural trade has drastically increased due to NAFTA, it has not absorbed displaced workers from corn production (estimated at around 600,000). Moreover, corn production has remained stable (at 20 million metric tons), arguably, as a result of income support to farmers, or a reluctance to abandon a millenarian tradition in Mexico: not only have peasants grown corn for millennia, corn originated in Mexico. Even today, Mexico is still the fourth largest corn producer in the world.

Potatoes

The area dedicated to potatoes has changed little since 1980 and average yields have almost tripled since 1961. Production reached a record 1.7 million tonnes in 2003. Per capita consumption of potato in Mexico stands at 17 kg a year, very low compared to its maize intake of 400 kg. On average, potato farms in Mexico are larger than those devoted to more basic food crops. Potato production in Mexico is mostly for commercial purposes; the production for household consumption is very small.

Sugar Cane

Approximately 160,000 small-and medium-sized farmers grow sugar cane in 15 Mexican states; currently there are 54 sugar mills around the country that produced 4.96 million tons of sugar in the 2009 crop, compared to 5.8 million tons in 2005. Mexico's sugar industry is characterized by high production costs and lack of investment. Mexico produces more sugar than it consumes. Sugar cane is grown on 700.000 ha in Mexico with a yield of 72 metric tons per ha.

Industry

The industrial sector as a whole has benefited from trade liberalization; in 2000 it accounted for almost 90% of all export earnings.

Among the most important industrial manufacturers in Mexico is the automotive industry, whose standards of quality are internationally recognized. The automobile sector in Mexico differs from that in other Latin American countries and developing nations in that it does not function as a mere assembly manufacturer. The industry produces technologically complex components and engages in some research and development activities. The "Big Three" (General Motors, Ford and Chrysler) have been operating in Mexico since the 1930s, while Volkswagen and Nissan built their plants in the 1960s. Later, Toyota, Honda, BMW, and Mercedes-Benz joined in. Given the high requirements of North American components in the industry, many European and Asian parts suppliers have also moved to Mexico: in Puebla alone, 70 industrial part-makers cluster around Volkswagen. The relatively small domestic car industry still is represented by DINA Camiones S.A. de C.V., that has built buses and trucks for almost half a century; Vehizero that builds hybrid trucks and the new car company Mastretta design that builds the Mastretta MXT sports car.

Some large industries of Mexico include Cemex, the third largest cement conglomerate in the world; the alcohol beverage industries, including world-renowned players like Grupo

Modelo; conglomerates like FEMSA, which apart from owning breweries and the OXXO convenience store chain, is also the second-largest Coca-Cola bottler in the world; Gruma, the largest producer of corn flour and tortillas in the world; and Grupo Bimbo, Telmex, Televisa, among many others. In 2005, according to the World Bank, high-tech industrial production represented 19.6% of total exports.

Maquiladoras (Mexican factories which take in imported raw materials and produce goods for export) have become the landmark of trade in Mexico. This sector has benefited from NAFTA, in that real income in the maquiladora sector has increased 15.5% since 1994, though from the non-maquiladora sector has grown much faster. Contrary to popular belief, this should be no surprise since maquiladora's products could enter the US duty free since the 1960s industry agreement. Other sectors now benefit from the free trade agreement, and the share of exports from non-border states has increased in the last 5 years while the share of exports from maquiladora-border states has decreased.

Currently Mexico is focusing in developing an aerospace industry and the assembly of helicopter and regional jet aircraft fuselages is taking place. Foreign firms such as MD Helicopters and Bombardier build helicopter and regional jets fuselages respectively in Mexico. Although the Mexican aircraft industry is mostly foreign, as is its car industry, Mexican firms have been founded such as Aeromarmi, which builds light propeller airplanes, and Hydra Technologies, which builds Unmanned Aerial Vehicles such as the S4 Ehecatl.

As compared with the United States or countries in Western Europe a larger sector of Mexico's industrial economy is food manufacturing which includes several world class companies but the regional industry is undeveloped. There are national brands that have become international and local Mom and Pop producers but little manufacturing in between.

Energy and Mineral Resources

Mineral resources are the "nation's property" (i.e. public property) by constitution. As such, the energy sector is administered by the government with varying degrees of private investment. Mexico is the sixth-largest oil producer in the world, with 3.7 million barrels per day. Pemex, the public company in charge of administering research, exploration and sales of oil, is the largest company (oil or otherwise) in Mexico, and the second largest in Latin America after Brazil's Petrobras.. Nonetheless, the company is heavily taxed, a significant source of revenue for the government, of almost 62 per cent of the company's sales. Without enough money to continue investing in finding new sources or upgrading infrastructure, and being protected constitutionally from private and foreign investment, some have predicted the company may face institutional collapse. While the oil industry is still relevant for the government's budget, its importance in GDP and exports has steadily fallen since the 1980s. In 1980 oil exports accounted for 61.6% of total exports; by 2000 it was only 7.3%.

Overview

The tertiary sector was estimated to account for 70.5% of the country's GDP, and employs 58% of the active population. This section includes transportation, commerce, warehousing, restaurant and hotels, arts and entertainment, health, education, financial and banking services, telecommunications as well as public administration and defence. Mexico's service sector is strong, and in 2001 replaced Brazil's as the largest service sector in Latin America in dollar terms.

Tourism

Tourism is one of the most important industries in Mexico. It is the fourth largest source of foreign exchange for the country. Mexico is the eighth most visited country in the world (with over 20 million tourists a year).

Financial Sector

Banking system: According to the IMF the Mexican banking system is strong, in which private banks are profitable and well-capitalized. The financial and banking sector is increasingly dominated by foreign companies or mergers of foreign and Mexican companies with the notable exception of Banorte. The acquisition of Banamex, one of the oldest surviving financial institutions in Mexico, by Citigroup was the largest US-Mexico corporate merger, at US $12.5 billion. In spite of that, the largest financial institution in Mexico is Bancomer associated to the Spanish BBVA. The process of institution building in the financial sector in Mexico has evolved hand in hand with the efforts of financial liberalization and of inserting the economy more fully into world markets. Over the recent years, there has been a wave of acquisitions by foreign institutions such as US-based Citigroup, Spain's BBVA and the UK's HSBC. Their presence, along with a better regulatory framework, has allowed Mexico's banking system to recover from the 1994–95 peso devaluation. Lending to the public and private sector is increasing and so is activity in the areas of insurance, leasing and mortgages. However, bank credit accounts for only 22% of GDP, which is significantly low compared to 70% in Chile. Credit to the Agricultural sector has fallen 45.5% in six years (2001 to 2007), and now represents about 1% of total bank loans. Other important institutions include savings and loans, credit unions, government development banks, "non-bank banks", bonded warehouses, bonding companies and foreign-exchange firms.

A wave of acquisitions has left Mexico's financial sector in foreign hands. Their foreign-run affiliates compete with independent financial firms operating as commercial banks, brokerage and securities houses, insurance companies, retirement-fund administrators, mutual funds, and leasing companies. Other important institutions include savings and loans, credit unions, government development banks, "non-bank banks", bonded warehouses, bonding companies and foreign-exchange firms.

Securities Market

Mexico has a single securities market, the Mexican Stock Exchange (*Bolsa Mexicana de Valores*, known as the Bolsa). The market has grown steadily, with its main indices increasing by more than 150% in 2003–05. It is Latin America's second largest exchange, after Brazil's. Still, the Bolsa remains relatively small when compared to other North American exchanges. The New York Stock Exchange is about 100 times larger; the Toronto Stock Exchange is six times larger.

The *Indice de Precios y Cotizaciones* (IPC, the general equities index) is the benchmark stock index on the Bolsa. In 2005 the IPC surged 37.8%, to 17,802.71 from 12,917.88, backed by a stronger Mexican economy and lower interest rates. It continued its steep rise through the beginning of 2006, reaching 19,272.63 points at end-March 2006. The stockmarket also posted a record low vacancy rate, according to the central bank. Local stockmarket capitalisation totalled US$236bn at end-2005, up from US$170bn at end-2004. As of March 2006 there were 135 listed companies, down from 153 a year earlier. Only a handful of the listed companies are foreign. Most are from Mexico City or Monterrey; companies from these two cities compose 67% of the total listed companies.

The IPC consists of a sample of 35 shares weighted according to their market capitalisation. Heavy hitters are America Telecom, the holding company that manages Latin America's largest mobile company, America Movil; Telefonos de Mexico, Mexico's largest telephone company; Grupo Bimbo, Mexico and Latin America's biggest baker; and Wal-Mart de Mexico, a subsidiary of the US retail giant. The makeup of the IPC is adjusted every six months, with selection aimed at including the most liquid shares in terms of value, volume and number of trades.

Mexico's stockmarket is closely linked to developments in the US. Thus, volatility in the New York and Nasdaq stock exchanges, as well as interest-rate changes and economic expectations in the US, can steer the performance of Mexican equities. This is both because of Mexico's economic dependence on the US and the high volume of trading in Mexican equities through American Depositary Receipts (ADRs). Currently, the decline in the value of the dollar is making non-US markets, including Mexico's, more attractive.

Despite the recent gains, investors remain wary of making placements in second-tier initial public offerings (IPOs). Purchasers of new issues were disappointed after prices fell in numerous medium-sized companies that made offerings in 1996 and 1997. IPO activity in Mexico remains tepid and the market for second-tier IPOs is barely visible. There were three IPOs in 2005.

Government Policies and the Central Bank

Banco de Mexico is Mexico's central bank, an internally autonomous public institution whose governor is appointed by the president and approved by the legislature to which it is fully responsible. Banco de Mexico's functions are outlined in the 28th article of the constitution and further expanded in the Monetary Law of the United Mexican States. Banco de Mexico's main objective is to achieve stability in the purchasing power of the national currency. It is also the lender of last resort.

Currency Policy

The floating exchange originated with reforms initiated after the December 1994 peso crash which had followed an unsustainable adherence to a short band. Under the new system, Banco de Mexico now makes no commitment to the level of the peso exchange rate, although it does employ an automatic mechanism to accumulate foreign reserves. It also possesses tools aimed at smoothing out volatility. The Exchange Rate Commission sets policy; it is made up of six members—three each from the Ministry of Finance and Public Credit (Secretaría de Hacienda y Credito Publico—SHCP) and the central bank, with the SHCP holding the deciding vote.

In August 1996, Banco de Mexico initiated a mechanism to acquire foreign reserves when the peso is strong, without giving the market signals about a target range for the exchange rate. The resulting high levels of reserves, mostly from petroleum revenues, have helped to improve the terms and conditions on debt Mexico places on foreign markets. However, there is concern that the government relies too heavily on oil income in order to build a healthy base of reserves. According to the central bank, international reserves stood at US $75.8 billion in 2007. In May 2003, Banco de Mexico launched a program that sells U.S. dollars via a monthly auction, with the goal of maintaining a stable, but moderate, level of reserves.

From April 1, 1998 through April 1, 2008 the Peso traded around a range varying from $8.46 MXN on per $1.00 USD on April 21, 1998 to $11.69 MXN per $1.00 USD on May 11, 2004, a 10 year peak depreciation of 38.18% between the two reference date extremes before recovering.

After the onset of the US credit crisis that accelerated in October 2008, the Peso had an exchange rate during October 1, 2008 through April 1, 2009 fluctuating from lowest to highest between $10.96 MXN per $1.00 USD on October 1, 2008 to

$15.42 MXN per $1.00 USD on March 9, 2009, a peak depreciation of 28.92% during those six months between the two reference date extremes before recovering.

From the $11.69 rate during 2004's low to the $15.42 rate during 2009's low, the peso depreciated 31.91% in that span covering the US recession coinciding Iraq War of 2003 and 2004 to the US & Global Credit Crisis of 2008.

Some experts including analysts at Goldman Sachs who coined the term BRIC in reference to the growing economics of Brazil, Russia, India, and China for marketing purposes believe that Mexico is going to be the 5th or 6th biggest economy in the world by the year 2050, behind China, United States, India, Brazil, and possibly Russia.

Monetary System

Mexico's monetary policy was revised following the 1994–95 financial crisis, when officials decided that maintaining general price stability was the best way to contribute to the sustained growth of employment and economic activity. As a result, Banco de Mexico has as its primary objective maintaining stability in the purchasing power of the peso. It sets an inflation target, which requires it to establish corresponding quantitative targets for the growth of the monetary base and for the expansion of net domestic credit.

The central bank also monitors the evolution of several economic indicators, such as the exchange rate, differences between observed and projected inflation, the results of surveys on the public and specialists' inflation expectations, revisions on collective employment contracts, producer prices, and the balances of the current and capital accounts.

A debate continues over whether Mexico should switch to a US-style interest rate-targeting system. Government officials in favour of a change say that the new system would give them more control over interest rates, which are becoming more important as consumer credit levels rise.

Until 2008, Mexico used a unique system, amongst the OECD countries, to control inflation in a mechanism known as the corto (lit. "shortage") a mechanism that allowed the central bank to influence market interest rates by leaving the banking system short of its daily demand for money by a predetermined amount. If the central bank wanted to push interest rates higher, it increased the corto. If it wished to lower interest rates, it decreased the corto. Source: BANXICO: in April 2004, the Central Bank began setting a referential overnight interest rate as its monetary policy.

Trade

Mexico is an export oriented economy. It is an important trade power as measured by the value of merchandise traded, and the country with the greatest number of free trade agreements. In 2005, Mexico was the world's fifteenth largest merchandise exporter and twelfth largest merchandise importer with a 12% annual percentage increase in overall trade. In fact, from 1991 to 2005 Mexican trade increased fivefold. Mexico is the biggest exporter and importer in Latin America; in 2005, Mexico alone exported US $213.7 billion, roughly equivalent to the sum of the exports of Brazil, Argentina, Venezuela, Uruguay, and Paraguay. However, Mexican trade is fully integrated with that of its North American partners: close to 90% of Mexican exports and 50% of its imports are traded with the United States and Canada. Nonetheless, NAFTA has not produced trade diversion. While trade with the United States increased 183% from 1993–2002, and that with Canada 165%, other trade agreements have shown even more impressive results: trade with Chile increased 285%, with Costa Rica 528% and Honduras 420%. Trade with the European Union increased 105% over the same time period.

Free Trade Agreements

Mexico joined the General Agreement on Tariffs and Trade (GATT) in 1986, and today is an active and constructive participant of the World Trade Organization. Fox's administration promoted the establishment of a Free Trade Area of the Americas; Puebla served as temporary headquarters for the negotiations, and several other cities are now candidates for its permanent headquarters if the agreement is reached and implemented.

Mexico has signed 12 free trade agreements with 44 countries:

- the North American Free Trade Agreement (NAFTA) (1994) with the United States and Canada;

- *Grupo de los tres,* Group of the three [countries], *or G-3 (1995) with Colombia and Venezuela; the latter decided to terminate the agreement in 2006; Mexico announced its intention to invite Ecuador, Peru or Panama as a replacement;*
- Free Trade Agreement with Costa Rica (1995);
- Free Trade Agreement with Bolivia (1995);
- Free Trade Agreement with Nicaragua (1998);
- Free Trade Agreement with Chile (1999);
- Free Trade Agreement with the European Union (2000);
- Free Trade Agreement with Israel (2000);
- TN Free Trade Agreement (2001), with Guatemala, El Salvador and Honduras;
- Free Trade Agreement with the European Free Trade Association (EFTA), integrated by Iceland, Norway, Liechtenstein and Switzerland (2001);
- Free Trade Agreement with Uruguay (2004); and
- Free Trade Agreement with Japan (2005).

Mexico has shown interest in becoming an associate member of Mercosur. The Mexican government has also started negotiations with South Korea, Singapore and Peru, and also wishes to start negotiations with Australia for a trade agreement between the two countries.

NAFTA

The North American Trade Agreement (NAFTA) is by far the most important Trade Agreement Mexico has signed both in the magnitude of reciprocal trade with its partners as well as in its scope. Unlike the rest of the Free Trade Agreements that Mexico has signed, NAFTA is more comprehensive in its scope and was complemented by the North American Agreement for Environmental Cooperation (NAAEC) and the North American Agreement on Labour Cooperation (NAALC).

The NAAEC agreement was a response to environmentalists' concerns that companies would relocate to Mexico or the United States would lower its standards if the three countries did not achieve a unanimous regulation on the environment. The NAAEC, in an aim to be more than a set of environmental regulations, established the North American Commission for Environmental Cooperation (NACEC), a mechanism for addressing trade and environmental issues, the North American Development Bank (NAD Bank) for assisting and financing investments in pollution reduction and the Border Environmental Cooperation Commission (BECC). The NAD Bank and the BECC have provided economic benefits to Mexico by financing 36 projects, mostly in the water sector. By complementing NAFTA with the NAAEC, it has been labelled the "greenest" trade agreement.

The NAALC supplement to NAFTA aimed to create a foundation for cooperation among the three members for the resolution of labour problems, as well as to promote greater cooperation among trade unions and social organizations in all three countries, in order to fight for the improvement of labour conditions. Though most economists agree that it is difficult to assess the direct impact of the NAALC, it is agreed that there has been a convergence of labour standards in North America. Given its limitations, however, NAALC has not produced (and in fact was not intended to achieve) convergence in employment, productivity and salary trend in North America.

The agreement fell short in liberalizing movement of people across the three countries. In a limited way, however, immigration of skilled Mexican and Canadian workers to the United States was permitted under the TN status. NAFTA allows for a wide list of professions, most of which require at least a Bachelor's degree, for which a Mexican or a Canadian citizen can request TN status and temporarily immigrate to the United States. Unlike the visas available to other countries, TN status requires no sponsorship, but simply a job offer letter.

The overall benefits of NAFTA have been quantified by several economists, whose findings have been reported in several publications like the World Bank's Lessons from NAFTA for LA and the Caribbean, NAFTA's Impact on North America, and NAFTA revisited by the Institute for International Economics. They assess that NAFTA has been positive for Mexico, whose poverty rates have fallen, and real income salaries have risen even after accounting for the 1994–1995 Economic Crisis. Nonetheless, they also state that it has not been enough, or fast enough, to produce an economic convergence nor to reduce the poverty rates substantially or to promote higher rates of growth. Some have suggested that in order to fully benefit from the agreement Mexico should invest in education and promote innovation as well as in infrastructure and agriculture.

Contrary to popular belief, the maquiladora program was in place far before NAFTA, in some sense dating all the way back to 1965. A maquiladora manufacturer operates by importing raw materials into Mexico either tariff free (NAFTA) or at a reduced rate on a temporary basis (18 months) and then using Mexico's relatively less expensive labour costs to produce finished goods for export. Prior to NAFTA maquiladora companies importing raw materials from anywhere in the world were given preferential tariff rates by the Mexican government so long as the finished good was for export. The US, prior to NAFTA, allowed Maquiladora manufactured goods to be imported into the US with the tariff rate only being applied to the value of non US raw materials used to produce the good, thus reducing the tariff relative to other countries. NAFTA has eliminated all tariffs on goods between the two countries, but for the maquiladora industry significantly increased the tariff rates for goods sourced outside of NAFTA.

Given the overall size of trade between Mexico and the United States, there are remarkably few trade disputes, involving relatively small dollar amounts. These disputes are generally settled in WTO or NAFTA panels or through negotiations between the two countries. The most significant areas of friction involve trucking, sugar, high fructose corn syrup, and a number of other agricultural products.

Mexican Trade Facilitation and Competitiveness

A research brief published by the World Bank as part of its Trade Costs and Facilitation Project suggests that Mexico has the potential to substantially increase trade flows and economic growth through trade facilitation reform. The study examines the potential impacts of trade facilitation reforms in four areas: port efficiency, customs administration, information technology, and regulatory environment (including standards).

The study projects overall increments from domestic reforms to be on the order of $31.8 billion, equivalent to 22.4 percent of total Mexican manufacturing exports for 2000-03. On the imports side, the corresponding figures are $17.1 billion and 11.2 percent, respectively. Increases in exports, including textiles, would result primarily from improvements in port efficiency and the regulatory environment. Exports of transport equipment would be expected to increase by the greatest increment from improvements in port efficiency, whereas exports of food and machinery would largely be the result of improvements in the regulatory environment. On the imports side, Mexican improvements in port efficiency would appear to be the most important factor, although for imports of transport equipment, improvements in service sector infrastructure would also be of relative importance.

Industry

Among the most important industrial manufacturers in Mexico is the automotive industry, whose standards of quality are internationally recognized. The automobile sector in Mexico differs from that in other Latin American countries and developing nations in that it does not function as a mere assembly manufacturer. The industry produces technologically complex components and engages in some research and development activities. The "Big Three" (General Motors, Ford and Chrysler) have been operating in Mexico since the 1930s, while Volkswagen and Nissan built their plants in the 1960s.

Later, Toyota, Honda, BMW, and Mercedes-Benz established a presence. Given the high requirements of North American components in the industry, many European and Asian parts suppliers have also moved to Mexico: in Puebla alone, 70 industrial part-makers cluster around Volkswagen. The relatively small domestic car industry is represented by DINA Camiones S.A. de C.V., which has built buses and trucks for almost half a century, and the new Mastretta company that builds the high performance Mastretta MXT sports car.

Some large industries of Mexico include Cemex, the third largest cement conglomerate in the world; the alcohol beverage industries, including world-renowned players like Grupo Modelo; conglomerates like FEMSA, which apart from owning breweries and the OXXO convenience store chain, is also the second-largest Coca-Cola bottler in the world; Gruma, the largest producer of corn flour and tortillas in the world; and Grupo Bimbo, Telmex, Televisa, among many others. In 2005, according to the World Bank, high-tech industrial production represented 19.6% of total exports.

Maquiladoras (Mexican factories which take in imported raw materials and produce goods for export) have become the landmark of trade in Mexico. This sector has benefited from NAFTA, in that real income in the maquiladora sector has increased 15.5% since 1994, though from the non-maquiladora sector has grown much faster. Contrary to popular belief, this should be no surprise since maquiladora's products could enter the US duty free since the 1960s industry agreement. Other sectors now benefit from the free trade agreement, and the share of exports from non-border states has increased in the last 5 years while the share of exports from maquiladora-border states has decreased.

Currently Mexico is focusing in developing an aerospace industry and the assembly of helicopter and commercial jet aircraft is taking place. Foreign firms such as MD Helicopters and Bombardier build helicopters and commercial jets respectively in Mexico. Although the Mexican aircraft industry is mostly foreign, as is its car industry, Mexican firms have been founded such as Aeromarmi, which builds light propeller airplanes, and Hydra Technologies, which builds Unmanned Aerial Vehicles.

As compared with the United States or countries in Western Europe, a larger sector of Mexico's industrial economy is food manufacturing, which includes several world class companies; but the regional industry is undeveloped. There are national brands that have become international, and local Mom and Pop producers, but little manufacturing in between.

Tourism

According to the World Tourism Organization, Mexico has one of the largest tourism industries in the world. In 2005 it was the seventh most popular. The most notable tourist draws are the ancient Mesoamerican ruins, and popular beach resorts. The coastal climate and unique culture – a fusion of European (particularly Spanish) and Mesoamerican cultures; also make Mexico attractive. The peak tourist seasons in Mexico are during December and during July and August, with brief surges during the week before Easter and during spring break at many of the beach resort sites which are popular among vacationing college students from the United States.

Mexico is the twenty-third highest tourism spender in the world, and the highest in Latin America.

Energy

Energy production in Mexico is managed by state-owned companies: the Federal Commission of Electricity (Comisión Federal de Electricidad, CFE) and Pemex (Petroleos Mexicanos). The CFE is in charge of the operation of electricity-generating plants and its distribution all across the territory, with the exception of the states of Morelos, Mexico State, Hidalgo and Mexico City, whose distribution of electricity is in charge of the state-owned Luz y Fuerza del Centro. Most of the electricity is generated in thermoelectrical plants, even though CFE operates several hydroelectric plants, as well as wind power, geothermal and nuclear generators.

Natural resources are the "nation's property" (i.e. public property) by constitution. As such, the oil sector is administered by the government with varying degrees of private investment. Mexico is the sixth-largest oil producer in the world, with 3.7 million barrels per day.

Pemex, the public company in charge of exploration, extraction, transportation and marketing of crude oil and natural gas, as well as the refining and distribution of petroleum products and petrochemicals, is one of the largest companies (oil or otherwise) in Latin America, making US $86 billion in sales a year, a sum larger than the GDP of some countries. Nonetheless, the company is heavily taxed, a significant source of revenue for the government, of almost 62 per cent of the company's sales. In 1980 oil exports accounted for 61.6% of total exports; by 2000 it was only 7.3%.

Transportation

As the third largest and second most populous country in Latin America, Mexico has developed an extensive transportation network to meet the needs of the economy. As with communications, transportation in Mexico is regulated by the Secretariat of Communications and Transportation, (Secretaria de Comunicaciones y Transporte, SCT) a federal executive cabinet ministry. Founded on 13 May 1891 as the Secretariat of Communications and Public Works, the SCT is divided into three subsecretariats: the Subsecretariat of Infrastructure, the Subsecretariat of Communications and the Subsecretariat of Transportation.

Roadways

The roadway network in Mexico is extensive and all areas in the country are covered by it. The roadway network in Mexico has an extension of 332,031 km (206,314 miles) of which 116,802 (72,577.398 miles) are paved, making it the largest paved-roadway network in Latin America. Of these, 10,474 km (6,508 miles) are multi-lane expressways: 9,544 km (5,930 miles) are four-lane highways and the rest have 6 or more lanes.

The highway network in Mexico is classified by number of lanes and type of access. The great majority of the network is composed of undivided or divided two-lane highways—with or without shoulders, and are known simply as carreteras. Four or more-lane freeways or expressways, with restricted or unrestricted access are known as autopistas. Speed limits in two-lane highways can vary depending on terrain conditions. The speed limit in multi-lane freeways or expressways is on average 110 km/h (70 mph) for automobiles and 95 km/h (60 mph) for buses and trucks.

The expressways are, for the most part, toll roads or autopistas de cuota. (Non-toll roads are referred to as carreteras libres, or free-roads). Most toll expressways have emergency telephone booths, water wells and emergency braking ramps at short intervals. The toll usually includes a "travellers' insurance" (seguro del viajero) should an accident occur within the freeway. Nonetheless, tolls are, on average, amongst the most expensive in the world, according to a comparative study realized in 2004 by the Chamber of Deputies. The most travelled freeways are those that link the three most populous cities in Mexico—Mexico City, Guadalajara and Monterrey—in the form of a triangle. No federal freeway or expressway crosses a city; toll expressways are either turned into toll bypasses (libramientos) often used as toll or free ring roads (perifericos), or are transformed into major arterial roads, even if they are, in function, freeways with restricted access.

Mexican highways are assigned a one to three-digit number. North-south highways are assigned odd numbers whereas east-west highways are assigned even numbers. Toll expressways usually run parallel to a free road, and therefore, are assigned the same number with the letter "D" added. (For example, the undivided two-lane highway connecting Mexico City and Puebla is MX 150, whereas the six-lane toll expressway is MX 150D).

Mexico has had difficulty in building an integrated highway network due to the country's orography and landscape characteristics—most of the country is crossed by high-altitude ranges of mountains. Over the last two decades, Mexico has made impressive investments in order to improve its road infrastructure and connect main cities and towns across the country. In spite of its extension and recent development, the roadway network in Mexico is still inadequate to meet the current needs of the population and, except for the toll roads, they are often not adequately maintained.

An additional problem is that in the centre of the country the roads run into Metro Mexico City from regional centers and there are few roads which run peripherically so as to connect the other regional centers without running through the congestion around the capital. The federal government (in parternship with the government of Mexico State and the Federal District) is trying to fix this problem by constructing a Mexico City bypass, named "Arco Norte"; it is willing to open it by the end of 2008.

Airports and Air Travel

Mexico has an extensive network of modern airports all throughout the territory; flying domestically is considered efficient and safe. Airport infrastructure in Mexico is the most advanced in Latin America: all the cities with more than 500,000 inhabitants have an airport. There are 1834 airports in Mexico, the third-largest number of airports by country in the world. The seven largest airports—which absorb 90% of air travel—are (in order of air traffic): Mexico City International Airport, Cancún International Airport, Don Miguel Hidalgo y Costilla

International Airport (Guadalajara), General Mariano Escobedo International Airport (Monterrey), General Abelardo L. Rodriguez International Airport (Tijuana), General Juan N. Alvares International Airport (Acapulco), and Lic. Gustavo Diaz Ordaz International Airport (Puerto Vallarta]). All airports are privately owned, with the exception of Mexico City International Airport. This airport remains the largest in Latin America and the 44th largest in the world transporting close to 26 million passengers a year.

There are more than 70 domestic airline companies in Mexico. The major players in the industry are Aeroméxico and Mexicana de Aviacion, the first is owned by Grupo Financiero Banamex and the second owned by Grupo Posadas. Other small airlines include Aeromexico Connect (Aeromexico regional subsidiary), Click Mexicana (Mexicana's low cost subsidiary), Aviacsa, Volaris, Interjet, Aeromar, Viva Aerobus, Magnicharters and Republicair.

The governments of the United States and Mexico recently approved an agreement of "open skies", which allows low-cost carriers to operate point-to-point (direct) routes between American and Mexican cities. This will decentralize air traffic in North America by bypassing major hubs and connecting smaller cities directly.

Seaports

Mexico has 76 seaports and 10 riverports. The four major seaports concentrating around 60% of the merchandise traffic are Altamira and Veracruz in the Gulf of Mexico, and Manzanillo and Lazaro Cardenas in the Pacific Ocean. These four seaports are followed in traffic by Acapulco, Puerto Vallarta, Guaymas, Tampico, Topolobambo, Mazatlan and Tuxpan.

The paved-roadway network in Mexico is the most extensive in Latin America at 116,802 km (72,577 mi) in 2005; 10,474 km (6,508 mi) were multi-lane freeways or expressways, most of which were tollways. Nonetheless, Mexico's diverse orography—most of the territory is crossed by high-altitude ranges of mountains—as well as economic challenges have led to difficulties in creating an integrated transportation network and even though the network has improved, it still cannot meet national needs adequately.

Being one of the first Latin American countries to promote railway development, the network, though extensive at 30,952 km (19,233 mi), is still inefficient to meet the economic demands of transportation. Most of the rail network is mainly used for merchandise or industrial freight and was mostly operated by National Railway of Mexico (Ferrocarriles Nacionales de Mexico, FNM), privatized in 1997.

In 1999, Mexico had 1,806 airports, of which 233 had paved runways; of these, 35 carry 97% of the passenger traffic. The Mexico City International Airport remains the largest in Latin America and the 44th largest in the world transporting 21 million passengers a year. There are more than 30 domestic airline companies of which only two are known internationally: Aeroméxico and Mexicana.

Mass transit in Mexico is modest. Most of the domestic passenger transport needs are served by an extensive bus network with several dozen companies operating by regions. Train passenger transportation between cities is limited. Inner-city rail mass transit is available at Mexico City—with the operation of the metro, elevated and ground train, as well as a Suburban Train connecting the adjacent municipalities of Greater Mexico City—as well as at Guadalajara and Monterrey, the first served by a commuter rail and the second by an underground and elevated metro.

Communications

The telecommunications industry is mostly dominated by Telmex (Telefonos de Mexico), privatized in 1990. As of 2006, Telmex had expanded its operations to Colombia, Peru, Chile, Argentina, Brazil and Uruguay and the United States. Other players in the domestic industry are Axtel and Maxcom. Due to Mexican orography, providing landline telephone service at remote mountainous areas is expensive, and the penetration of line-phones per capita is low compared to other Latin American countries, at forty-percent, however 82% of Mexicans over the age of 15 own a mobile phone. Mobile telephony has the advantage of reaching all areas at a lower cost, and the total number of mobile lines is almost two times that of landlines, with an estimation of 63 million lines. The telecommunication industry is regulated by the government through Cofetel

(Comision Federal de Tele-communications). The Mexican satellite system is domestic and operates 120 earth stations. There is also extensive microwave radio relay network and considerable use of fiber-optic and coaxial cable. Mexican satellites are operated by Satelites Mexicanos (Satmex), a private company, leader in Latin America and servicing both North and South America. It offers broadcast, telephone and telecommunication services to 37 countries in the Americas, from Canada to Argentina. Through business partnerships Satmex provides high-speed connectivity to ISPs and Digital Broadcast Services. Satmex maintains its own satellite fleet with most of the fleet being Mexican designed and built.

Mexico has recently emerged as a major producer of communications technology. In 2008 Mexico manufactured over 130 million mobile phones making it the sixth largest producer of mobile phones.

Usage of radio, television, and Internet in Mexico is prevalent. There are approximately 1,410 radio broadcast stations and 236 television stations (excluding repeaters). Major players in the broadcasting industry are Televisa—the largest Spanish media company in the Spanish-speaking world—and TV Azteca.

Demographics

According to the latest official estimate, which reported a population of 111 million, Mexico is the most populous Spanish-speaking country in the world. Mexican annual population growth has drastically decreased from a peak of 3.5% in 1965 to 0.99% in 2005. Life expectancy in 2006 was estimated to be at 75.4 years (72.6 male and 78.3 female). The states with the highest life expectancy are Baja California (75.9 years) and Nuevo Leon (75.6 years). The Federal District has a life expectancy of the same level as Baja California.

The lowest levels are found in Chiapas (72.9), Oaxaca (73.2) and Guerrero (73.2 years). The mortality rate in 1970 was 9.7 per 1000 people; by 2001, the rate had dropped to 4.9 men per 1000 men and 3.8 women per 1000 women. The most common reasons for death in 2001 were heart problems (14.6% for men 17.6% for women) and cancer (11% for men and 15.8% for women).

Mexican population is increasingly urban, with close to 75% living in cities. The five largest urban areas in Mexico (Greater Mexico City, Greater Guadalajara, Greater Monterrey, Greater Puebla and Greater Toluca) are home to 30% of the country's population. Migration patterns within the country show positive migration to north-western and south-eastern states, and a negative rate of migration for the Federal District. While the annual population growth is still positive, the national net migration rate is negative (-4.7/1000), attributable to the emigration phenomenon of people from rural communities to the United States.

Metropolitan Areas

Metropolitan areas in Mexico have been traditionally defined as the group of municipalities that heavily interact with each other, usually around a core city. In 2004, a joint effort between CONAPO, INEGI and the Ministry of Social Development (SEDESOL) agreed to define metropolitan areas as either:

- the group of two or more municipalities in which a city with a population of at least 50,000 is located whose urban area extends over the limit of the municipality that originally contained the core city incorporating either physically or under its area of direct influence other adjacent predominantly urban municipalities all of which have a high degree of social and economic integration or are relevant for urban politics and administration; or
- a single municipality in which a city of a population of at least one million is located and fully contained, (that is, it does not transcend the limits of a single municipality); or
- a city with a population of at least 250,000 which forms a conurbation with other cities in the United States.

It should be noted, however, that northwestern and southeastern states are divided into a small number of large municipalities whereas central states are divided into a large number of smaller municipalities. As such, metropolitan areas in the northwest usually do not extend over more than one municipality (and figures usually report population for the entire municipality) whereas metropolitan areas in the centre extend over many municipalities. Few metropolitan areas extend beyond the limits of one state, namely: Greater Mexico City (Federal District, Mexico State and Hidalgo), Puebla-Tlaxcala (Puebla and Tlaxcala, but excludes the city of Tlaxcala), Comarca Lagunera (Coahuila and Durango), and Tampico (Tamaulipas and Veracruz).

Immigration

Mexico is home to the largest number of U.S. citizens abroad (estimated at one million as of 1999), which represents 1% of the Mexican population and 25% of all U.S. citizens abroad. Other significant communities of foreigners are those of Central and South America, most notably from Argentina, Brazil, Chile, Colombia, Peru, Cuba, Venezuela, Guatemala, and Belize. Though estimations vary, the Argentine community is considered to be the second largest foreign community in the country (estimated somewhere between 30,000 and 150,000). Mexico also received a large number of Lebanese. The Mexican-Lebanese community now numbers around 400,000.

Throughout the 20th century, Mexico followed a policy of granting asylum to fellow Latin Americans and Europeans (mostly Spaniards in the 1940s) fleeing political persecution in their home countries. In October 2008, Mexico tightened its immigration rules and agreed to deport Cubans using the country as an entry point to the US. Because Mexico is much richer than the countries to its immediate southeast, it has a chronic problem with illegal immigration from those countries, especially Guatemala, Honduras, and El Salvador. Large numbers of Central American migrants who have crossed Guatemala's western border into Mexico are deported every year.

Discrepancies between the figures for official legal aliens and those of all foreign-born residents regardless of their immigration status are quite large. The official figure for foreign-born legal residents in Mexico is 493,000 (since 2004), with a majority (86.9%) of these born in the United States (except Chiapas, where the majority of immigrants are from Central America). The five states with the most immigrants are Baja California (12.1% of total immigrants), Mexico City (the Federal District; 11.4%), Jalisco (9.9%), Chihuahua (9%) and Tamaulipas

(7.3%). More than 54.6% of the immigrant population are fifteen years old or younger, while 9% are fifty or older. Illegal immigration has been a problem for Mexico, especially since the 1970s. In 2006 Mexico detained more than 182,000 people who entered the country illegally, mainly from nearby Guatemala, Honduras, El Salvador, all being Central-American countries neighbouring Mexico to the south. Smaller numbers of illegal immigrants come from Ecuador, Cuba, China, South Africa, and Pakistan.. Mexico represents the largest source of immigration to the United States. About 9% of the population born in Mexico is now living in the United States. 28.3 million Americans listed their ancestry as Mexican as of 2006.

Ethnic Groups

Mexico is ethnically diverse, and the constitution defines the country to be a multicultural nation. Mexican nationality is relatively young, stemming back only to 1821 when Mexico achieved independence from the Spanish empire, and it consists of many, separate regional and ethnic groups such as the various indigenous peoples and European immigrants. The majority of Mexicans are Mestizos which makes up the core of the Mexican cultural identity.

In 2004, the Mexican government founded the National Institute of Genomic Medicine (INMEGEN) which launched the Mexican Genome Diversity Project. In May 2009, the Institute issued a report on a major genomic study of the Mexican population. Among the findings, it was reported that of the 80% of the population that is mestizo, the proportions of European and indigenous ancestry are approximately even. The proportions of admixture were found to vary geographically from north to south, as previous pre-genomic studies had surmised, with the European contribution predominating in the north and the indigenous component greater in the south. One of the significant conclusions of the study as reported was that even while it is composed of diverse ancestral genetic groups from around the world, the Mexican population is genetically distinctive among the world's populations.

Language

There is no de jure constitutional official language at the federal level in Mexico. Spanish, spoken by 97% of the population, is considered a national language by The General Law of Linguistic Rights of the Indigenous Peoples, which also grants all indigenous minority languages spoken in Mexico, regardless of the number of speakers, the same validity as Spanish in all territories in which they are spoken, and indigenous peoples are entitled to request some public services and documents in their native languages.

Mexican law has granted these indigenous minority languages the status of "national languages", along with Spanish. The law includes all Amerindian languages regardless of origin; that is, it includes the Amerindian languages of ethnic groups non-native to the territory. As such the National Commission for the Development of Indigenous Peoples recognizes the language of the Kickapoo, which immigrated from the United States, and recognizes the languages of the Guatemalan Amerindian refugees. The Mexican government has promoted and established intercultural bilingual primary and secondary education in some indigenous rural communities. Approximately 5.4% of the population speaks an indigenous language and 1.2% do not speak Spanish.

Mexico has the largest Spanish-speaking population in the world with more than twice as many as the second largest Spanish-speaking country. (Spain, Argentina, and Colombia all have about 40 million speakers each.) Almost a third of all Spanish native speakers in the world live in Mexico. Nahuatl is spoken by 1.5 million people and Yucatec Maya by 800,000. Some of the national languages are in danger of extinction; Lacandon is spoken by fewer than one hundred people.

English is widely used in business at the border cities, as well as by the one million U.S. citizens that live in Mexico, mostly retirees in small towns in Baja California, Guanajuato and Chiapas. There are some 80,000 German-speaking Mennonites in Mexico.

Religion

Mexico has no official religion, and the government does not provide any financial contributions to the church, and the church does not participate in public education.

The last census reported, by self-ascription, that 95% of the population is Christian. Roman Catholics are 89% of the total population, 47% percent of whom attend church services weekly. In absolute terms, Mexico has the world's second largest number of Catholics after Brazil.

About 6% of the population (more than 4.4 million people) is Protestant, of whom Pentecostals and Charismatics (called Neo-Pentecostals in the census) are the largest group (1.37 million people). There are also a sizeable number of Seventh-day Adventists (0.6 million people). The 2000 national census counted more than one million Jehovah's Witnesses. The Church of Jesus Christ of Latter-day Saints claims over one million registered members as of 2009. About 25% of registered members attend a weekly sacrament service although this can fluctuate up and down. There are eleven Temples of the Church of Jesus Christ of Latter-day Saints in Mexico.

The presence of Jews in Mexico dates back to 1521, when Hernan Cortés conquered the Aztecs, accompanied by several Conversos. According to the last national census by the INEGI, there are now more than 45,000 Mexican Jews. Almost three million people in the 2000 National Census reported having no religion. Islam in Mexico is practiced by 25,000 people in Mexico. It is practiced in populations such as in the city of Torreon, Coahuila, and there are an estimated 300 Muslims in the San Cristobal de las Casas area in Chiapas. Mexico's Buddhist population currently makes up a tiny minority, some 108,000 according to latest accounts. Most of its members are of Asian descent, while people of various other walks of life have turned toward Buddhism in the recent past.

In 1992, Mexico lifted almost all restrictions on the Catholic Church and other religions, including granting all religious groups legal status, conceding them limited property rights, and lifting restrictions on the number of priests in the country. Until recently, priests did not have the right to vote, and even now they cannot be elected to public office.

Culture

Mexican culture reflects the complexity of the country's history through the blending of pre-Hispanic civilizations and the culture of Spain, imparted during Spain's 300-year colonization of Mexico. Exogenous cultural elements mainly from the United States have been incorporated into Mexican culture. The Porfirian era (el Porfiriato), in the last quarter of the nineteenth century and the first decade of the twentieth century, was marked by economic progress and peace. After four decades of civil unrest and war, Mexico saw the development of philosophy and the arts, promoted by President Diaz himself. Since that time, as accentuated during the Mexican Revolution, cultural identity has had its foundation in the mestizaje, of which the indigenous (i.e. Amerindian) element is the core. In light of the various ethnicities that formed the Mexican people, Jose Vasconcelos in his publication La Raza Comica (The Cosmic Race) (1925) defined Mexico to be the melting pot of all races (thus extending the definition of the mestizo) not only biologically but culturally as well. This exalting of mestizaje was a revolutionary idea that sharply contrasted with the idea of a superior pure race prevalent in Europe at the time.

Cinema

Mexican films from the Golden Age in the 1940s and 1950s are the greatest examples of Latin American cinema, with a huge industry comparable to the Hollywood of those years. Mexican films were exported and exhibited in all of Latin America and Europe. Maria Candelaria (1944) by Emilio Fernandez, was one of the first films awarded a Palme d'Or at the Cannes Film Festival in 1946, the first time the event was held after World War II. The famous Spanish-born director Luis Bunuel realized in Mexico, between 1947 to 1965 some of him master pieces like Los olvidados (1949), Viridiana (1961) and El angel exterminador (1963). Famous actors and actresses from this period include Maria Felix, Pedro Infante, Dolores del Rio, Jorge Negrete and the comedian Cantinflas.

More recently, films such as Como agua para chocolate (1992), Cronos (1993), Amores perros (2000), Y tu mama tambien (2001), El Crimen del Padre Amaro (2002), Pan's Labyrinth (2006) and Babel (2006) have been successful in creating universal stories about contemporary subjects, and were internationally recognised, as in the prestigious Cannes Film Festival. Mexican directors Alejandro Gonzalez Inarritu (Amores perros, Babel), Alfonso Cuaron (Children of Men, Harry Potter and the Prisoner of Azkaban), Guillermo del Toro, Carlos Carrera (The Crime of Father Amaro), and screenwriter

Guillermo Arriaga are some of the most known present-day film makers.

Music

Mexican society enjoys a vast array of music genres, showing the diversity of Mexican culture. Traditional music includes Mariachi, Banda, Norteno, Ranchera and Corridos; on an everyday basis most Mexicans listen to contemporary music such as pop, rock, etc. in both English and Spanish. Mexico has the largest media industry in Latin America, producing Mexican artists who are famous in Central and South America and parts of Europe, especially Spain. Some well-known Mexican singers are Thalia, Luis Miguel, Alejandro Fernandez, Julieta Venegas and Paulina Rubio. Popular groups are Cafe Tacuba, Molotov and Mana, among others.

According to the Sistema Nacional de Fomento Musical, there are between 120 and 140 youth orchestras affiliated to this federal agency from all federal states. Some states, through their state agencies in charge of culture and the arts—Ministry or Secretariat or Institute or Council of Culture, in some cases Secretariat of Education or the State University—sponsor the activities of a professional Symphony Orchestra or Philharmonic Orchestra so all citizens can have access to this artistic expression from the field of classical music. There is no public information about the exact number of professional orchestras in the country (probably 40 ensembles of very diverse caliber). Mexico City is the most intense hub of this activity hosting 12 professional orchestras sponsored by different agencies such as the National Intitute of Fine Arts, the Secretariat of Culture of the Federal District, The National University, the National Polytechnic Institute, a Delegacion Political (Coyoacan) and very few are a kind of private ventures.

Orquestas in Mexico are mainly subsidized by a governmental body or agency, unlike their American counterparts, therefore, these organizations do not have departments such as marketing or development. States such as Baja California Sur, Campeche, Chiapas, Colima, Morelos, Nayarit, Quintana Roo, Sonora, Tabasco, and Tlaxcala do not have a professional Symphony orchestra. The only permanent opera company belongs to the National Institute of Fine Arts, offering six productions yearly, however, some cities such as Guadalajara, Monterrey or Morelia make important efforts to present this kind of expression to local audiences.

Fine Arts

Post-revolutionary art in Mexico had its expression in the works of renowned artists such as Frida Kahlo, Diego Rivera, Jose Clemente Orozco, Rufino Tamayo, Federico Cantu Garza, David Alfaro Siqueiros and Juan O'Gorman. Diego Rivera, the most well-known figure of Mexican Muralism, painted the Man at the Crossroads at the Rockefeller Centre in New York City, a huge mural that was destroyed the next year due to the inclusion of a portrait of Russian communist leader Lenin. Some of Rivera's murals are displayed at the Mexican National Palace and the Palace of Fine Arts.

Academic music composers of Mexico include Manuel Maria Ponce, Jose Pablo Moncayo, Julian Carrillo, Mario Lavista, Carlos Chavez, Silvestre Revueltas, Arturo Marquez, and Juventino Rosas, many of whom incorporated traditional elements into their music. Nobel Prize winner Octavio Paz, Carlos Fuentes, Juan Rulfo, Elena Poniatowska, and Jose Emilio Pacheco, are some of the most recognized authors of Mexican literature.

Broadcast Media

Two of the major television networks based in Mexico are Televisa and TV Azteca. Televisa is also the largest producer of Spanish-language content in the world and also the world's largest Spanish-language media network. Grupo Multimedios is another media conglomerate with Spanish-language broadcasting in Mexico, Spain, and the United States. Soap operas (telenovelas) are translated to many languages and seen all over the world with renowned names like Veronica Castro, Lucia Mendez, Lucero, and Thalia. Even Gael Garcia Bernal and Diego Luna from Y tu mama tambien and current Zegna model have appeared in some of them.

Some of their TV shows are modelled after counterparts from the U.S. like Family Feud (100 Mexicanos Dijeron or "A hundred Mexicans said" in Spanish) and Que dice la gente, Big Brother, American Idol, Saturday Night Live and others.

Nationwide news shows like Las Noticias por Adela on Televisa resemble a hybrid between Donahue and Nightline. Local news shows are modelled after counterparts from the U.S. like the Eyewitness News and Action News formats. Border cities receive television and radio stations from the U.S., while satellite and cable subscription is common for the middle-classes in most cities, and they often watch movies and TV shows from the U.S.

Cuisine

Mexican cuisine is known for its intense and varied flavours, colourful decoration, and variety of spices. Most of today's Mexican food is based on pre-Columbian traditions, including the Aztecs and Maya, combined with culinary trends introduced by Spanish colonists.

The conquistadores eventually combined their imported diet of rice, beef, pork, chicken, wine, garlic and onions with the native pre-Columbian food, including maize, tomato, vanilla, avocado, papaya, pineapple, chili pepper, beans, squash, limes (limon in Mexican Spanish), sweet potato, peanut and turkey.

Mexican food varies by region, because of local climate and geography and ethnic differences among the indigenous inhabitants and because these different populations were influenced by the Spaniards in varying degrees. The north of Mexico is known for its beef, goat and ostrich production and meat dishes, in particular the well-known Arrachera cut.

Central Mexico's cuisine is largely made up of influences from the rest of the country, but also has its authentics, such as barbacoa, pozole, menudo, tamales, and carnitas.

Southeastern Mexico, on the other hand, is known for its spicy vegetable and chicken-based dishes. The cuisine of Southeastern Mexico also has quite a bit of Caribbean influence, given its geographical location. Seafood is commonly prepared in the states that border the Pacific Ocean or the Gulf of Mexico, the latter having a famous reputation for its fish dishes, a la veracruzana.

In modern times, other cuisines of the world have become very popular in Mexico, thus adopting a Mexican fusion. For example, sushi in Mexico is often made with a variety of sauces based on mango or tamarind, and very often served with serrano-chili-blended soy sauce, or complemented with vinegar, habanero and chipotle peppers. The most internationally recognized dishes include chocolate, tacos, quesadillas, enchiladas, burritos, tamales and mole among others. Regional dishes include mole poblano, chiles en nogada and chalupas from Puebla; cabrito and machaca from Monterrey, cochinita pibil from Yucatan, Tlayudas from Oaxaca, as well as barbacoa, chilaquiles, milanesas, and many others.

Sports

Mexico City hosted the XIX Olympic Games in 1968, making it the first Latin American city to do so (Rio de Janeiro will be the second in 2016). The country has also hosted the FIFA World Cup twice, in 1970 and 1986.

Mexico's most popular sport is association football (soccer). It is commonly believed that Football was introduced in Mexico by Cornish miners at the end of the 19th century. By 1902 a five-team league had emerged with a strong English influence. Football became a professional sport in 1943. Since the "Era Professional" started, Mexico's top clubs have been Guadalajara with 11 championships, America with 10 and Toluca with 9. In Mexican Football many players have been raised to the level of legend, but two of them have received international recognition above others. Antonio Carbajal was the first player to appear in five World Cups, and Hugo Sanchez was named best CONCACAF player of the 20th century by IFFHS. Mexican's biggest stadiums are Estadio Azteca, Estadio Olimpico Universitario and Jalisco Stadium. Notable achievements by Mexican national soccer teams include winning the 1999 Confederations Cup, the 2005 U-17 World Cup, and being runners-up in the 1977 U-20 World Cup, 1993 Copa America, and 2001 Copa America. The Mexican team was also a runner-up in the 2007 Beach Soccer World Cup. Mexican football clubs have achieved good results in international competitions like the Copa Libertadores and the FIFA Club World Championship.

The national sport of Mexico is Charreada. Bullfighting is also a popular sport in the country, and almost all large cities have bullrings. Plaza Mexico in Mexico City, is the largest bullring in the world, which seats 55,000 people. Professional wrestling (or Lucha libre in Spanish) is a major crowd draw with national promotions such as AAA, LLL, CMLL and others.

Baseball is very popular, especially in the Gulf of Mexico, Yucatan Peninsula and the Northern States. In some regions baseball has traditionally been more popular than soccer. The season runs from March to July with play-offs held in August. The Mexican professional league is named the Liga Mexicana de Beisbol. Current champions (2007) are Sultanes de Monterrey who defeated in a tight series Leones de Yucatan. However, the best level of baseball is played in Liga Mexicana del Pacifico, played in Sinaloa, Sonora and Baja California. Given that it is played during the MLB off-season, some of its players are signed to play with the league 8 teams. Current champions (2007) are Yaquis de Obregon. The league champion participates in the Caribbean Series, a tournament between the Champions of Winter Leagues of Mexico, Venezuela, Puerto Rico and the Dominican Republic; the 2009 Caribbean Series edition will be held in Mexicali. While usually not as strong as the United States, the Caribbean countries and Japan, Mexico has nonetheless achieved several international baseball titles. Mexico has had several players signed by Major League teams, the most famous of them being Dodgers pitcher Fernando Valenzuela.

Mexican clubs (such as Monterrey La Raza) participate in North American indoor soccer leagues. Mexico is also a leading country in Basque Pelota. Tennis has also traditionally been popular in Mexico, having produced some great players like Rafael Osuna, though international successes have been very few in recent years.

Mexico is an international power in professional boxing (at the amateur level, several Olympic boxing medals have also been won by Mexico). Vicente Saldivar, Ruben Olivares, Salvador Sanchez, Julio Cesar Chavez, Ricardo Lopez and Erik Morales are but a few Mexican fighters who have been ranked among the best of all time.

The most important professional basketball league is the Liga Nacional de Baloncesto Professional and covers the whole Mexican territory, where the Soles de Mexicali are the current champions. In 2007 three Mexican teams will be competing in the American Basketball Association. In the northwestern states is the CIBACOPA Competition, with professional basketball players from Mexico and the U.S. Universities and some teams from the NBA. Mexico is a major international power in Taekwondo. Mexican athletes have achieved renown in disciplines like marathon running, racewalking and diving.

American football is played at the major universities like ITESM, UANL, UDLA, IPN and UNAM. The college league in Mexico is called ONEFA. Several Mexican players have been signed by the NFL over the years. Rugby is played at the amateur level throughout the country with the majority of clubs in Mexico City and others in Monterrey, Guadalajara, Celaya, Guanajuato and Oaxaca.

Auto racing is very popular in Mexico. Throughout the years, Mexico has hosted races for some of the most important international championships such as Formula One, NASCAR, Champ Car, A1 Grand Prix, among others. Mexico also has its own NASCAR-sanctioned stock car series, the NASCAR Mexico, which runs 14 events in different cities, drawing large crowds. Other forms of racing include Formula Renault, Formula Vee, touring cars, Pick-up trucks, endurance racing, rallying, and off-road.

Ice hockey is played in larger cities like Monterrey, Guadalajara, Villahermosa, Culiacán and Mexico City.

Notable Mexican athletes include golfer Lorena Ochoa, who is currently ranked first in the LPGA world rankings, Ana Guevara, former world champion of the 400 metres (1,300 ft) and Olympic subchampion in Athens 2004, and Fernando Platas, a numerous Olympic medal winning diver.

Sport fishing is popular in Baja California and the big Pacific coast resorts, while freshwater bass fishing is growing in popularity too. The gentler arts of diving and snorkeling are big around the Caribbean, with famous dive sites at Cozumel and on the reefs further south. The Pacific coast is becoming something of a centre for surfing, with few facilities as yet; all these sports attract tourists to Mexico.

Another sport in which Mexico is competitive at the international level is polo.

Healthcare and Education

Since the early 1990s, Mexico entered a transitional stage in the health of its population and some indicators such as mortality patterns are identical to those found in highly developed countries like Germany or Japan. Although all Mexicans are entitled to receive medical care by the state, 50.3 million Mexicans had no medical insurance as of 2002. Efforts to increase the number of people are being made, and the current administration intends to achieve universal health care by 2011.

Mexico's medical infrastructure is highly rated for the most part and is usually excellent in major cities, but rural areas and indigenous communities still lack equipment for advanced medical procedures, forcing patients in those locations to travel to the closest urban areas to get specialized medical care.

State-funded institutions such as Mexican Social Security Institute (IMSS) and the Institute for Social Security and Services for State Workers (ISSSTE) play a major role in health and social security. Private health services are also very important and account for 13% of all medical units in the country.

Medical training is done mostly at public universities with some specializations done abroad. Some public universities in Mexico, such as the University of Guadalajara, have signed agreements with the U.S. to receive and train American students in Medicine. Health care costs in private institutions and prescription drugs in Mexico are on average lower than that of its North American economic partners.

In 2004, the literacy rate was at 97% for youth under the age of 14 and 91% for people over 15, placing Mexico at the 24th place in the world rank accordingly to UNESCO. Primary and secondary education (9 years) is free and mandatory. Even though different bilingual education programs have existed since the 1960s for the indigenous communities, after a constitutional reform in the late 1990s, these programs have had a new thrust, and free text books are produced in more than a dozen indigenous languages.

In the 1970s, Mexico established a system of "distance-learning" through satellite communications to reach otherwise inaccessible small rural and indigenous communities. Schools that use this system are known as telesecundarias in Mexico. The Mexican distance learning secondary education is also transmitted to some Central American countries and to Colombia, and it is used in some southern regions of the United States as a method of bilingual education. There are approximately 30,000 telesecundarias and approximately a million telesecundaria students in the country.

The largest and most prestigious public university in Mexico, today numbering over 269,000 students, is the National Autonomous University of Mexico (Universidad Nacional Autónoma de Mexico, UNAM) founded in 1910. Three Nobel laureates and most of Mexico's modern-day presidents are among its former students. UNAM conducts an astounding 50% of Mexico's scientific research and has presence all across the country with satellite campuses and research centers. The National Autonomous University of Mexico ranks 15th place in the Top 200 World University Ranking published by The Times Higher Education Supplement in 2008, making it the highest ranked Spanish-speaking university in the world and the highest ranked in Latin America. The second largest university is the National Polytechnic Institute (IPN). These institutions are public, and there are at least a couple of public universities per state.

One of the most prestigious private universities is Monterrey Institute of Technology and Higher Education (ITESM). It was ranked by the Wall Street Journal as the 7th top International Business School worldwide and 74th among the world's top arts and humanities universities ranking of The Times Higher Education Supplement, published in 2005. ITESM has thirty-two secondary campuses, apart from its Monterrey Campus. Other important private universities include Mexico's Autonomous Technological Institute (ITAM), ranked as the best economics school in Latin America, Fundación Universidad de las Americas, Puebla (UDLAP) and the Ibero-American University (Universidad Iberoamericana).

In April 2009 Mexico saw the outbreak of a new strain of H1N1 influenza, which has infected up to 800 people and infected over 140,000 worldwide as of July 20.

Science and Technology

After the Viceroyalty of New Spain was founded, the Spanish brought the scientific culture that dominated Spain to the

Viceroyalty of New Spain. The municipal government (cabildo) of Mexico City formally requested the Spanish crown to establish a university in 1539. The Royal and Pontifical University of Mexico (Real y Pontificia Universidad de Mexico) was established in 1551.

The university was administered by the clergy and it was the official university of the empire. It provided quality education for the people, and it was a hub of intellectual and religious development in the region.

It taught subjects such as physics and mathematics from the perspective of Aristotelian philosophy. Augustinian philosopher Alonso Gutierrez wrote Physica speculatio, America's first scientific text, in 1557. By the late eighteenth century, the university had trained 1,162 doctors, 29,882 bachelors, and many lawyers.

During the Mexican Enlightenment, science can be divided into the four period: the early period (from 1735 to 1767), the creole period (from 1768 to 1788), the official or Spanish period (from 1789 to 1803), and the period of synthesis (from 1804 to the beginning of the Mexico's independence movement in 1810).

During the sixteenth and seventeenth centuries, modern science developed in Europe, but it lagged behind in Mexico. The new ideas developed in science in Europe were not important in Mexico. The 1767 expulsion of Jesuits, who had introduced the new ideas in Mexico, helped to antagonize the creoles, and also promoted nation feelings among Mexicans.

After the expulsion, self-taught upper class civilians were the first scientists in Mexico. Later on, they were joined by the Spanish scientists, and they did research, teaching, publishing, and translating texts. The ideas of Francis Bacon and René Descartes were freely discussed at seminars, which caused scholasticism to lose strength. During the Mexican Enlightenment, Mexico made progress in science. Progress were made in subjects such as astronomy, engineering, etc. In 1792 the Seminary of Mining was established. Later it became College of Mining, in which the first modern physics laboratory in Mexico was established.

Famous scientists of the Enlightenment included Jose Antonio de Alzate y Ramirez and Andres Manuel del Rio. Rio discovered the chemical element vanadium in 1801.

The Mexican War of Independence brought temporary end to Mexico's scientific progress. The Royal and Pontifical University of Mexico closed in 1833. For many years, there were no scientific activities in Mexico. The Royal and Pontifical University of Mexico was definitively shut down in 1865.

During the end of the nineteenth century, the process of industrialization began in Mexico. Under the influence of positivists and scientific thinkers, the government assisted in public education. In 1867 Gabino Barreda, a student of Auguste Comte, was charged with the commission aimed at reforming education.

Subjects such as physics, chemistry, and mathematics were included into the secondary school curriculum. National Preparatory School was established. The influence of positivists led to a renaissance of scientific activity in Mexico. General Manuel Mondragon, invented the first automatic rifle the Mondragon rifle during this time.

During the twentieth century, Mexico made significant progress in science and technology. New universities and research institutes were established. The National Autonomous University of Mexico (Universidad Nacional Autónoma de Mexico, UNAM) was officially established in 1910, and the university become one of the most important institutes of higher learning in Mexico. UNAM provides world class education in science, medicine, and engineering. Many scientific institutes and new institutes of higher learning, such as National Polytechnic Institute (founded in 1936), were established during the first half of the twentieth century. Most of the new research institutes were created within UNAM. Twelve institutes were integrated into UNAM from 1929 to 1973.

In 1959, the Mexican Academy of Sciences (Academia Mexicana de Ciencias) was established as a non-governmental, non-profit organization of distinguished scientists. The Academy has grown in membership and influence, and it represents a strong voice of scientists from different fields, mainly in science policy.

By 1960, science was institutionalized in Mexico. It was viewed as a legitimate endeavour by the Mexican society.

In 1960, the Centre for Research and Advanced Studies of the National Polytechnic Institute was established as a centre for graduate studies in subjects such as biology, mathematics, and physics. In 1961, the institute began its graduate programs in physics and mathematics and schools of science were established in Mexican states of Puebla, San Luis Potosi, Monterrey, Veracruz, and Michoacan. The Academy for Scientific Research was established in 1969 and the National Council of Science and Technology was established in 1971.

In 1985 Rodolfo Neri Vela became the first Mexican citizen to enter space as part of the STS-61-B mission.

In 1995 Mexican chemist Mario J. Molina shared the Nobel Prize in Chemistry with Paul J. Crutzen, and F. Sherwood Rowland for their work in atmospheric chemistry, particularly concerning the formation and decomposition of ozone. Molina, an alumnus of UNAM, became the first Mexican citizen to win the Nobel Prize in science.

In recent years, the biggest scientific project being developed in Mexico was the construction of the Large Millimeter Telescope (Gran Telescopio Milimetrico, GMT), the world's largest and most sensitive single-aperture telescope in its frequency range. It was designed to observe regions of space obscured by stellar dust.

In 1962, the National Commission of Outer Space (Comisión Nacional del Espacio Exterior, CONNE) was established, but was dismantled in 1977. In 2007, a project was presented to reopen a new Mexican Space Agency (AEXA) and it was approved at the end of 2008 with the headquarters set to be located in the state of Hidalgo.

Government institutions such as SEMAR and SEDENA have also developed advanced microprocessors, imaging systems, military A.I. systems, rockets, software, long range ballistic missiles, electronic devices and electronic military subsystems many of which have been sold to other Latin American nations. Other consumer electronics companies such as Mabe have been functioning since the nineteen fifties and have expanded out of Latin America into markets around the world such as Asia and Europe and even into the United States where a large percentage of American branded appliances are actually of Mexican design and origin but sold under local brand names. In fact as of 2008 one out of every four consumer appliances sold in the United States was of Mexican origin.

Based on the information managed by the Scopus, a bibliographic database for science, the Spanish web portal SCImago places Mexico in the position 18 of the country scientific ranking with 82,792 publications, and in the position 34 if considering its value of 134 for the h-index. Both positions are computed for the period 1996-2.

7

Geography of Latin America

Introduction

South America is the southern continent of the Americas, situated in the Western Hemisphere and mostly in the Southern Hemisphere, with a relatively small portion in the Northern Hemisphere. It is bordered on the west by the Pacific Ocean and on the north and east by the Atlantic Ocean; North America and the Caribbean Sea lie to the northwest.

America was named in 1507 by cartographers Martin Waldseemuller and Matthias Ringmann after Amerigo Vespucci, who was the first European to suggest that the lands newly discovered by Europeans were not India, but a New World unknown to Europeans.

South America has an area of 17,840,000 square kilometers (6,890,000 sq mi), or almost 3.5% of the Earth's surface. As of 2005, its population was estimated at more than 371,090,000. South America ranks fourth in area (after Asia, Africa, and North America) and fifth in population (after Asia, Africa, Europe, and North America).

South America occupies the southern portion of the landmass sometimes referred to as the New World. The continent is generally delimited on the northwest by the Darien watershed along the Colombia–Panama border, or (according to some sources) by the Panama Canal which transects the Isthmus of Panama. Geopolitically and geographically all of Panama – including the segment east of the Panama Canal in the isthmus – is typically included in North America alone and among the countries of Central America. Almost all of mainland South America sits on the South American Plate.

Many of the islands of the Caribbean (or West Indies) – e.g., the Leeward and Lesser Antilles – sit atop the Caribbean Plate, a tectonic plate with a diffuse topography. The islands of Aruba, Barbados, Trinidad, and Tobago sit on the northerly South American continental shelf. The Netherlands Antilles and the federal dependencies of Venezuela lie along the northerly South American shelf. Geopolitically, the island states and overseas territories of the Caribbean are generally grouped as a part or subregion of North America. The South American nations that border the Caribbean Sea—including Colombia, Venezuela, Guyana, Suriname, and French Guiana—are also known as Caribbean South America. Other islands are the Galapagos islands that belong to Ecuador and Easter Island (in Oceania but belongs to Chile), Robinson Crusoe Island, Chiloe, and the Tierra del Fuego.

South America is home to the world's highest waterfall, Angel Falls in Venezuela; the largest river (by volume), the Amazon River; the longest mountain range, the Andes (whose highest mountain is Aconcagua at 6,962 m; the driest place on earth, the Atacama Desert; the largest rainforest, the Amazon Rainforest; the highest capital city, La Paz, Bolivia; the highest commercially navigable lake in the world, Lake Titicaca; and, excluding research stations in Antarctica, the world's southernmost permanently inhabited community, Puerto Toro, Chile.

South America's major mineral resources are gold, silver, copper, iron ore, tin, and petroleum. The many resources of South America have brought high income to its countries especially in times of war or of rapid economic growth by industrialized countries elsewhere. However, the concentration in producing one major export commodity often has hindered the development of diversified economies. The fluctuation in the price of commodities in the international markets has led historically to major highs and lows in the economies of South American states, often causing extreme political instability. This is leading to efforts to diversify production to drive away from staying as economies dedicated to one major export. South America is one of the most biodiverse continents on earth.

South America is home to many interesting and unique species of animals including the llama, anaconda, piranha, jaguar, vicuna, and tapir. The Amazon rainforests possess high biodiversity, containing a major proportion of the Earth's species. Regions in South America include the Andean States, the Guianas, the Southern Cone, and Brazil which is the largest country by far, in both area and population.

The geography of South America contains many diverse regions and climates. Geographically, South America is generally considered a continent forming the southern portion of the American landmass, south and east of the Panama-Colombia border by most authorities, or south and east of the Panama Canal by some. South and North America are sometimes considered a single continent or supercontinent, while constituent regions are infrequently considered subcontinents. Geopolitically and geographically, all of Panama – including the segment east of the Panama Canal in the isthmus – is generally considered a part of North America alone and among the countries of Central America.

South America became attached to North America only recently (geologically speaking) with the formation of the Isthmus of Panama some 3 million years ago, which resulted in the Great American Interchange. The Andes, likewise a comparatively young and seismically restless mountain range, run down the western edge of the continent; the land to the east of the Andes is largely tropical rain forest, the vast Amazon River basin. The continent also contains drier regions such as eastern Patagonia and the extremely arid Atacama desert.

The South American continent also includes various islands, most of which belong to countries on the continent. The Caribbean territories are grouped with North America. The South American nations that border the Caribbean Sea — including Colombia, Venezuela, Guyana, Suriname, and French Guiana — are also known as Caribbean South America.

Topography and Geology

The geographical structure of South America is deceptively simple for a continent-sized landmass. The continent's topography is often likened to a huge bowl owing to its flat interior almost ringed by high mountains. With the exception of narrow coastal plains on the Pacific and Atlantic Oceans, there are three main topographic features: the Andes, a central lowland, and the extensive Brazilian and Guiana Highlands in the east.

The Andes are a Cenozoic mountain range formed (and still forming) by the continuing collision of the American and Pacific tectonic plates. In their northern and central reaches the Andes are quite wide and contain extensive plateaux such as the Altiplano and a number of major valleys such as the Rio Magdalena.

These contain three of the world's highest capitals: Bogota, Quito and highest of all, La Paz, Bolivia. The southern Andes have been eroded by the Patagonian Ice Sheet and are much lower and narrower. There are a number of large glaciers in the northern part, but from latitude 19°S to 28°S the climate is so arid that no permanent ice can form even on the highest peaks. Permafrost, however, is widespread in this section of the Altiplano and continuous above 5,600 metres (18,373 ft).

The climate of the coastal belt west of the Andes shows violent contrasts, including two of the world's wettest regions in the Colombian Choco and southern Chile and the world's driest desert, the Atacama. This dry area is cooled by the Humboldt Current and upwelling, giving rise to the largest fisheries in the world. There are two small transition zones between the perhumid and perarid regions: around Guayaquil with summer rain, and the Mediterranean climate region of central Chile. Both these regions have highly erratic rainfall strongly influenced by El Nino events, which bring major floods. In contrast, the high plateaux of the Andes are drier than normal during El Nino episodes.

The very fertile soils from the erosion of the Andes formed the basis for the continent's only pre-Columbian state civilizations: those of the Inca Empire and its predecessors (Chavin, Nazca, Mochica, etc.). The area is still a major agricultural region. The Altiplano contains many rare minerals such as copper, tin, mercury ore. The Atacama is mined for its nitrates. East of the Andes in Peru is regarded as the most important biodiversity hotspot in the world with its unique forests that form the western edge of the world's largest rainforest, the Amazon Rainforest.

East of the Andes is a large lowland drained by a small number of rivers, including the two largest in the world by drainage area-the Amazon River and the more southerly Parana River.

The other major river of this central lowland is the Orinoco River, which has a natural channel linking it with the Amazon. Most of this central lowland is sparsely populated because the soils are heavily leached, but in the south is the very fertile pampas of Argentina-one of the world's major food-producing regions where wheat and beef cattle are pre-eminent. The natural vegetation of the northern lowlands are either savanna in the northern *llanos* and southern *campos*, or tropical rainforest throughout most of the Amazon basin. Efforts to develop agriculture, outside of fertile floodplains of rivers descending from the Andes, have been largely failures because of the soils. Cattle have long been raised in the *llanos* of northern Colombia and Venezuela, but petroleum is now the dominant industry in the northern lowlands, making Venezuela the richest country in the continent.

The eastern highlands are much older than the Andes, being pre-Cambrian in origin, but are still rugged in places, especially in the wet tepuis of Venezuela, Guyana and Roraima. The Amazon River has cut a large valley through a former highland, and to the east is a relatively low plateau comprising the Nordeste and Southeast regions of Brazil.

In the north of this region is the arid *sertao*, a poor region consistently affected by extremely erratic rainfall, and the humid *zona da mata*, once home of the unique Atlantic Rainforest with many species not found in the Amazon, and now a centre for sugarcane. Further south, the main land use is coffee, while Sao Paulo is the economic heart of the continent with its industry.

South of about Santa Catarina, the highlands fade out to low plains in Uruguay.

East of the Andes in Argentina, there are a number of rugged, generally dry *lslands*, the highest of which is the Sierra de Cordoba near the city of that name. Argentine Patagonia is a Paleozoic plateau now heavily dissected by rivers flowing from the Andes.

Territories

The largest country in South America by far, in both area and population, is Brazil, followed by Argentina. Regions in South America include the Andean States, the Guianas, the Southern Cone, and Eastern South America.

The Physical Geography

Physical Geography of Latin America

Why study physical geography?

1) It's important in its own right. Many geographers spend all their time looking at physical features.
2) The natural environment also influences how people are able to live in a particular area. To understand fully the people of Latin America, we need to understand something about their environment.

What do I want you to take from this presentation?

Familiarity with what Latin America's physical environment looks like.

An idea of the major physical features and categorization systems. You do not need to remember the details. Think about summarizing the major points.

Location

Latin America extends from 32.3 degrees North (Northern Baja CA) to 55 degrees South (southern islands of Tierra del Fuego). Two major sub-regions of Latin America = Middle America and South America.

Three parts of Middle America = Mexico, Central America, Caribbean Islands.

South America

South America's physical geography can be summarized into three main regions: 1) the Andes, 2) the Eastern Highlands, and 3) the Lowlands.

The Andes: The most prominent feature of South America is the Andes, which stretch the length of the continent. The mountain system is narrow, averaging barely 200 miles in width. Note how close to the Pacific coast the Andes are. The continental divide (the line along the highest ridges, from which water flows west to the Pacific and east to the Atlantic) is thus very close to the Pacific coast.

What are the implications of this? How would Latin America be different if the Andes were on the East of the continent?- Much drier as the prevailing winds bring moisture from the East across the widest part of the continent. Rainbearing clouds would be forced to rise and lose their moisture very close to the East coast where they arrive, leaving much of the continent dry. This is what happens in Australia. The Andes are high, with the highest peak, Aconcagua in Chile, reaching nearly 23,000 feet. The passes throughout the Andes are high and narrow, steep, and dangerous. The majority of the Andes are volcanic, many of them still active. Geologically-speaking, they are relatively young mountains. The Andes were formed through tectonic instability. The mid-Atlantic Ridge is a divergent plate boundary which pushes South America Westwards. We will return to this idea later.

Western Slopes of Andes

The rivers flowing westwards toward the Pacific carry a limited amount of water, both because they are short and because the western slopes of the mountains where their watersheds are located are extremely dry as the prevailing weather systems come across Latin America from the East and therefore lose most of their water on the Eastern side of the Andes. The Western edge of the Andes is largely desert. However, small areas are relatively well-watered from the few rivers that do come down from the Andes. These "oases" have been cultivated for a long time and a variety of civilizations have developed in them, e.g. Moche Civilization of North Western Peru.

Eastern Slopes of Andes

On the Eastern slopes of the Andes, the rivers are fuller and longer, as they gather the abundant rainfall from the Eastern slopes of the Andes. At their start, they are tempestuous, fast-flowing streams. Further downstream, the gradients become gentler and the rivers slower-moving and wider. The main river flowing east is the Amazon. The Amazon's course is 4000 miles long; the river carries one-fifth of the world's freshwater and drains 40% of South America. The Amazon Basin receives on average 3m of rain per year. 45 million gallons of water are discharged into the Atlantic every second from the Amazon. Each year an area the size of England is flooded in the wet season. The gradient from the foothills of the Andes to the Eastern edge of the Amazon Basin is extremely gentle. Some parts of Peru are only 200m in elevation, for example. The Amazon has over 1000 major tributaries. 1000 miles before it reaches the sea the Amazon is already 10 miles wide. For much of its course it drops barely 2" a mile. There is an island within its delta that is larger than Switzerland.

Altiplano

Cold, thin soils, but flat land available, so has been a center of settlement for many years, e.g. Inca Empire, Uros people and their floating islands on Lake Titicaca.

The Eastern Highlands

In contrast to the Andes, the eastern highlands in South America are very ancient. They consist of three parts: the Brazilian Highlands, Guiana Highlands, and Patagonian Plateau. The Guiana and Brazilian Highlands are part of the same basic geological structure. They have been separated from each other through erosion by the Amazon River. The Patagonian Plateau is less obvious but forms the southernmost end of the eastern highland region. It is semi-arid and extends from the Rio Negro, at the southern edge of the pampas, southward for 1,000 miles to eastern Tierra del Fuego. Along their western margin these highlands are separated from the Andes by a north/south-running trough called the "Pre-Andean Depression." Erosion has worn the Eastern Highlands down to elevations that range from about 600 to 4,500 feet. From the perspective of people living in the Andes, the Eastern Highlands are not "highlands" at all. The Western edge of the Highlands is called the Mato Grosso Plateau. It is a mature, deeply-weathered landscape of rolling hills, interspersed with higher more resistant structures known locally as either serras (mountain ranges) or chapadas (plateaux). Instead, the eastern edge of the Brazilian Highlands is steep, although not terribly high. The eastern edge is also rich in minerals. The steep drop-off and spectacular waterfalls on the western edge of the Brazilian Highlands create tremendous potential for hydroelectric power. The mountains just behind Rio de Janeiro are part of the Great Escarpment, the eastern edge of the Brazilian Highlands which rises steeply from the Atlantic. The Great Escarpment has always made movement between the coast and the interior of Brazil difficult.

Plate Tectonics

To understand how the Eastern Highlands were formed, we have to consider plate tectonics. The idea that the continents were at one time joined was put forward by a German meteorologist, Alfred Wegener, in 1912. He noted the geologic "fit" between the bulge of South America and the dent in Africa and postulated that the two were once joined in a single landmass, which he called Pangaea. Subsequent evidence for the existence of prehistoric links between the continents were provided from several lines of enquiry. For instance:

The existence of similar plants and animals on the now separate continents, e.g. opposums in South America are the only marsupial mammals found outside of Australia, suggesting a link between South America and Australia at one time. Also, the same species of fossils have been found on now separate landmasses, e.g. South America and Africa. Distinctive rock strata – scientists have found that the geologic structure of the rocks of SW Africa and SE Brazil are identical in composition and age, suggesting that the two areas were once joined.

The idea that the continents have moved over the earth's surface has also been supported by various pieces of evidence:

Now-extinct plants and animals that were adapted to climates very different from today. Also, coal (which can only form in warm and wet conditions) is found under Antarctica. Evidence of sea-floor spreading added credence to the idea that ocean basins, and formations of the earth's crust are not static. Scientists subsequent to Wegener suggested that there was not one super-landmass (Pangaea), but actually two: a northern hemisphere mass (Laurasia) and a southern hemisphere mass (Gondwanaland). Gondwanaland is thought to have existed from around 300 to 200 million years ago when it broke up into pieces that began to move towards their current positions. The

Eastern Highlands of South America date back geologically to this time. When Gondwanaland broke apart into separate continents, there was a tremendous out-pouring of lava from beneath the earth's surface. The Parana basalt plateau, part of the Brazilian Highlands, is a remnant of this phenomenon. Plate tectonics continue to influence South America to this day. The South American Plate continues to move westward at the rate of a few centimetres per year, pushed by the divergence of the Mid-Atlantic Ridge. As it moves, it is bumping against and overriding the Nazca Plate, which is being submerged beneath the western edge of South America. This creates intense pressure, which is relieved by the uplifting of the Andes and periodic volcanic eruptions and earthquakes. Tectonic forces generated by this collision of the two Plates trigger volcanic eruptions and earthquakes and have lifted parts the Andes 5000m during the past 20 million years.

The Lowlands

The Llanos

The northernmost of the lowlands are the *llanos*. The lowlands called the *llanos* in South America are the grassy floodplain of the Orinoco River, in Venezuela and Colombia. The llanos occupy an earth depression that was formerly an extension of the Atlantic Ocean. Apart from some areas around oil reserves, the llanos are one of the least developed regions of Latin America.

Amazon Basin

Not all of the Amazon Basin is rainforested; some is also grassland, e.g., Beni, Bolivia. Although supporting apparently uniform tropical rainforest, the Amazon rainforest has a remarkable number of distinctive regions, affected by different micro-climates, elevations, soils, etc. e.g. friajes, flooded forest. The Amazon floods annually, inundating an area the size of England. As a result, the plants and animals of this region have to be adapted to floods.

Gran Chaco and Paraguay-Parana-La Plata River System

The Gran Chaco is a dry and forbidding region, with few

permanent streams and low agricultural potential. It is also a poverty-stricken region of South America. What water does leave the area is exported to the Atlantic via South America's third great river system (in addition to the Orinoco and Amazon Rivers and their tributaries)-the Paraguay-Parana-La Plata river system, which empties into the Atlantic between Uruguay and Argentina.

Agriculture is possible in this area, but tends to be small-scale and marginal owing to shifting river courses and extreme seasonal fluctuations in water courses. Too sparsely populated for permanent irrigation systems, so farmers tend to plant crops in shallow depressions, called *banyados*, whose locations shift from year to year. The upper course of the Paraguay River drains a great inland swamp known as the Pantanal, a very rich area for wildlife.

The Pampas

Bordering the impoverished Gran Chaco to the South is the Argentine Pampas, one of Latin America's most prosperous and productive agricultural regions. It is an alluvial plain on which fine-grained, wind-borne soil particles have been deposited (loess). These have been enriched over time by volcanic deposits blown from the Andes and rich organic matter from its grasses. Result has been the formation of a prime prairie soil. Topsoil is up to 1000 feet deep in places. Area also has adequate rainfall, moderate temperatures, and is about the size of Texas.

The Pampas has sustained much of Argentina's economic growth through ranching. This was only made possible in the late 1800s after the invention of barbed wire. The cow-boy image of Argentina comes from this area. Many native species now rare.

The Economic Geography

Economic geography is the study of the location, distribution and spatial organization of economic activities across the world. The subject matter investigated is strongly influenced by the researcher's methodological approach. Neoclassical location theorists, following in the tradition of Alfred Weber, tend to focus on industrial location and use quantitative methods. Since the 1970s, two broad reactions against neoclassical approaches have significantly changed the discipline: Marxist political economy, growing out of the work of David Harvey; and the new economic geography which takes into account social, cultural, and institutional factors in the spatial economy.

Economic geography is usually regarded as a subfield of the discipline of geography, although recently economists such as Paul Krugman and Jeffrey Sachs have pursued interests that can be considered part of economic geography. Krugman has gone so far as to call his application of spatial thinking to international trade theory the "new economic geography", which directly competes with an approach within the discipline of geography that is also called "new economic geography". The name geographical economics has been suggested as an alternative.

Given the variety of approaches, economic geography has taken to many different subject matters, including: the location of industries, economies of agglomeration (also known as "linkages"), transportation, international trade and development, real estate, gentrification, ethnic economies, gendered economies, core-periphery theory, the economics of urban form, the relationship between the environment and the economy (tying into a long history of geographers studying culture-environment interaction), and globalization. This list is by no means exhaustive.

The economy of South America comprises around 371 million people living in twelve nations and three territories.

Economic Development

As of early 2007, South America is experiencing great economic development, with Venezuela, Colombia, Argentina, Uruguay and Peru growing their economies by over 8% per annum. Brazil's economy, on the other hand, is expected to grow by a more sluggish pace during the year.

South America relies heavily on the exporting of goods. On an exchange rate basis Brazil (the seventh largest economy in the world and the second largest in the Americas) leads the way in total amount of exports at $137.8 billion dollars followed by Chile at 58.12 billion and Argentina with 46.46 billion.

GDP Ranks as of 2008

GDP (PPP) 2009

Note: French Guiana, unlisted here, had a total GDP of US$3.52 billion in 2006 (at real exchange rates, not at PPP).

Rank in World	*Country*	*GDP*
10	Brazil	$ 2,024,000 million
24	Argentina	$ 558,000 million
29	Colombia	$ 400,300 million
32	Venezuela	$ 355,200 million
45	Peru	$ 253,400 million
47	Chile	$ 244,300 million
65	Ecuador	$ 107,100 million
91	Bolivia	$ 45,130 million
92	Uruguay	$ 44,050 million
107	Paraguay	$ 28,280 million
163	Suriname	$ 4,182 million
172	Guyana	$ 2,973 million

Source: CIA World Factbook

GDP Per Capita (PPP) 2009

Note: French Guiana, unlisted here, had a GDP per capita of US$17,336 in 2006 (at real exchange rates, not at PPP).

Rank in world	*Country*	*GDP per capita*
57	Chile	$ 14,700
58	Argentina	$ 13,800
96	Venezuela	$ 13,200
65	Uruguay	$ 12,600
68	Brazil	$ 10,513
99	Colombia	$ 9,200
103	Suriname	$ 8,800
97	Peru	$ 9,107
101	Ecuador	$ 7,300
125	Bolivia	$ 4,600
107	Paraguay	$ 4,100
105	Guyana	$ 3,900

Source: List of countries by GDP (PPP) per capita

GNI Per Capita

Gross national income (GNI) is the sum of value added by all resident producers plus any product taxes (less subsidies) not included in the valuation of output plus net receipts of primary income (compensation of employees and property income) from abroad. GNI per capita is:

Country	*GNI per capita*
Brazil	$ 4,730
Suriname	$ 3,200
Peru	$ 2,920
Ecuador	$ 2,840
Colombia	$ 2,740
Paraguay	$ 1,400
Guyana	$ 1,130
Bolivia	$ 1,100

Source: World Bank

External Debt

External debt (or foreign debt) is that part of the total debt in a country that is owed to creditors outside the country. The debtors can be the government, corporations or private households. The debt includes money owed to private commercial banks, other governments, or international financial institutions such as the IMF and World Bank.

Rank in World	*Country*	*External Debt*
26	Brazil	$ 216.10 billion
33	Argentina	$ 118.00 billion
47	Chile	$ 49.18 billion
48	Venezuela	$ 45.44 billion
49	Colombia	$ 43.30 billion
63	Peru	$ 27.81 billion
72	Ecuador	$ 17.56 billion
78	Uruguay	$ 12.00 billion
108	Bolivia	$ 3.80 billion
111	Paraguay	$ 3.632 billion
151	Guyana	$ 1.20 billion (2002 est.)
164	Suriname	$ 504.30 million (2005 est.)

Source: CIA World Factbook

Annual Economic Growth

Country	*Annual Economic Growth (%)*
Peru	9.8%
Argentina	8.7%
Venezuela	8.4%
Colombia	7.0%
Uruguay	7.0%
Paraguay	6.4%
Ecuador	6.3%
Suriname	5.5%
Brazil	5.4%
Guyana	5.4%
Chile	5.0%
Bolivia	4.2%

Source: List of countries by GDP (real) growth rate

Unemployment Rate (Lowest to Highest)

Unemployment Rate	*(%)*
Paraguay	5.60
Peru	6.90
Chile	7.00
Argentina	7.3
Brazil	7.50
Bolivia	7.50
Venezuela	8.50
Uruguay	9.20
Ecuador	9.30
Suriname	9.50
Colombia	11.20
Guyana	N/A

Source: CIA World Factbook

Poverty Line (Lowest to Highest)

Year Ofestimate	*Country*	*Population below Poverty line (%)*
2009	Chile	13.7 (cepal)
2009	Argentina	15.3
2006	Uruguay	27.4
2006	Brazil	27.5
2005	Paraguay	32.0
2009	Peru	36.8
2005	Venezuela	37.9
2006	Ecuador	38.3
2005	Colombia	49.2
2006	Bolivia	60.0
2002	Suriname	70.0
N/A	Guyana	-

Source: CIA World Factbook

Annual Inflation

Country	*Annual Inflation(%)*
Peru	1.80%
Ecuador	2.20%
Brazil	3.60%
Chile	4.40%
Colombia	5.50%
Suriname	6.40%
Paraguay	8.10%
Uruguay	8.10%
Bolivia	8.70%
Argentina	8.80%
Guyana	12.20%
Venezuela	18.70%

Source: CIA World Factbook

Public Debt (Lowest to Highest)

Refers to what is owed by the combined public sector to both domestic and foreign creditors.

Country	*Public Debt (% of GDP)*
Chile	3.60
Venezuela	20.4
Paraguay	27.10
Peru	29.30
Ecuador	30.40
Brazil	41.50
Colombia	53.90
Bolivia	46.20
Uruguay	53.1
Argentina	59.00
Guyana	N/A
Suriname	N/A

Source: CIA World Factbook

Trade Blocks

The biggest Trade Block in South America is Mercosur (or Mercosul in Portuguese), comprising Argentina, Brazil, Paraguay, Uruguay and Venezuela. Associate states include Bolivia, Chile, Colombia, Ecuador and Peru. The second-biggest trade bloc is the Andean Community of Nations comprising Bolivia, Colombia, Ecuador, Peru, Venezuela and as of 2006 Chile.

The Union of South American Nations is expected to merge both trade blocs.

Economic Sectors

Agriculture

Main products include: Coffee, soybeans, wheat, rice, corn, sugarcane, cocoa, citrus, beef

Manufacturing

Industries are also important to South America's economy. Most South American factories produce food items, consumer goods, or building materials. Larger countries also produce cars, trucks, and airplanes. Some of these companies import all the parts and raw materials needed for manufacturing which limits the amount of profits they can receive for the item. An important factor that is crucial to the success of industries is importing and exporting. An organization called Mercosur helps to expand trade, improve transportation, and reduce tariffs among member countries.

Transport

At the beginning of August 2008, Venezuelan president Hugo Chavez and his colleagues from Argentina and Brazil spoke about Latin American integration and Chavez threw an ambitious idea out: a train that would connect Venezuela's capital (Caracas) with Argentina's (Buenos Aires), and cities in between.

8

Central America

Central America is the central geographic region of the Americas. It is the southernmost, isthmian portion of the North American continent, which connects with South America on the southeast. Central America is considered to be part of the Mesoamerican biodiversity hotspot, excluding the southern portions of Panama.

Geography

Central America has an area of 524,000 square kilometers (202,000 sq mi), or almost 0.1% of the Earth's surface. As of 2009, its population was estimated at 41,739,000. It has a density of 77 people per square kilometre or 206 people per square mile.

Physical Geography

Physiographically, Central America is the tapering isthmus of southern North America, extending from the Isthmus of Tehuantepec in southern Mexico southeastward to the Isthmus of Panama where it connects to the Colombian Pacific Lowlands in northwestern South America. Alternatively, the Trans-Mexican Volcanic Belt delimits the region on the north. Central America has an area of some 592,000 square kilometres. The Pacific Ocean lies to the southwest, the Caribbean Sea lies to the northeast, and the Gulf of Mexico lies to the north. Most of Central America rests atop the Caribbean Plate.

The region is geologically active, with volcanic eruptions and earthquakes occurring from time to time. In 1976 Guatemala was hit by a major earthquake, killing 23,000 people; Managua, the capital of Nicaragua, was devastated by earthquakes in 1931 and 1972, the last one killed about 5,000 people; three earthquakes devastated El Salvador, one in 1986 and two in 2001; one earthquake devastated northern and central Costa Rica in 2009 killing at least 34 people; in Honduras a powerful earthquake killed 7 people in 2009.

Volcanic eruptions are common in the region. In 1968 the Arenal Volcano, in Costa Rica, erupted and killed 87 people. Fertile soils from weathered volcanic lavas have made it possible to sustain dense populations in the agriculturally productive highland areas.

Central America has many mountain ranges; the longest are the Sierra Madre de Chiapas, the Cordillera Isabelia and the Cordillera de Talamanca. Between the mountain ranges lie fertile valleys that are suitable for the people; in fact most of the population of Honduras, Costa Rica and Guatemala live in valleys. Valleys are also suitable for the production of coffee, beans and other crops.

History

In pre-Columbian times, the north-western areas of modern Central America were part of the Mesoamerican civilization. The Native American societies of Mesoamerica occupied the land ranging from central Mexico in the north to Costa Rica in the south. Most notable among these were the Maya, who had built numerous cities throughout the region, and the Aztecs, who created a vast empire. The pre-Columbian cultures of Panama traded with both Mesoamerica and South America, and can be considered transitional between those two cultural areas.

Following Christopher Columbus's discovery of the Americas for Spain, the Spanish sent numerous expeditions to the region, and they began their conquest of Maya lands in the 1520s. In 1540, Spain established the Captaincy General of Guatemala, which extended from southern Mexico to Costa Rica, and thus encompassed most of what is currently known as Central America, with the exception of British Honduras (present-day Belize). This lasted nearly three centuries, until a rebellion (which followed closely on the heels of the Mexican War of Independence) in 1821.

After the dissolution of Spanish authority, the former Captaincy General remained intact as part of the short-lived First Mexican Empire, then turned into the Federal Republic of Central America, which was a representative democracy with its capital at Guatemala City. This union consisted of the present day nations of Guatemala (which included the former state of Los Altos), El Salvador, Honduras, Nicaragua, Costa Rica (which included a region which is now part of Panama, and the Guanacaste Province which was once part of Nicaragua), and Soconusco, a portion of the modern Mexican state of Chiapas. The Republic lasted from 1823 to 1838, when it began to disintegrate due to civil wars.

Central American Integration

Central America is going through a process of political, economic and cultural transformation that started in 1907 with the creation of the Central American Court of Justice. In 1951 the integration process continued with the signature of the San Salvador Treaty that created the ODECA, the Organization of Central American States. Unfortunately, the ODECA was not completely successful due to internal conflicts between several states of the region.

It was until 1991 that the integration agenda was completed with the creation of the SICA, *Sistema para la Integración Centroamericana* or System for the Central American Integration. The SICA provided a clear legal base to avoid discrepancies between the member states. The SICA membership includes the 7 nations of Central America plus the Dominican Republic, a state that is part of the Caribbean.

On December 6, 2008 SICA announced an agreement to pursue a common currency and common passport for the member nations. No timeline for implementation was discussed.

Central America already has several supranational institutions such as the Central American Parliament, the Central American Bank for Economic Integration and the Central American Common Market.

Foreign Relations

Until recently, all Central American countries have maintained diplomatic relations with the Republic of China

(Taiwan) instead of the People's Republic of China. President Oscar Arias of Costa Rica, however, established diplomatic relations with the People's Republic of China in 2007, severing formal diplomatic ties with the Republic of China (Taiwan).

Usage

"Central America" may mean different things to different people in the world according to the context:

In English speaking countries, Central America is considered a region of the North American continent. Geopolitically, it usually comprises seven countries – Belize, Guatemala, El Salvador, Honduras, Nicaragua, Costa Rica, and Panama. Mexico, in whole or in part, is sometimes included. Some geographers include the five states of Campeche, Chiapas, Tabasco, Quintana Roo, and Yucatan, together representing 12.1% of the country's total area.

In Latin America, Iberia, and some other parts of Europe, the Americas are considered to be a single continent, and Central America is considered a region of this continent. In Ibero-America, the region is defined as seven nations – Belize, Guatemala, El Salvador, Honduras, Nicaragua, Costa Rica, and Panama – and may occasionally include Mexico's southernmost region. Geopolitically, Mexico is considered part of North America.

Occasionally, the term *Central America* is used synonymously with *Middle America*. Among some German geographers, *Mittelamerika* may be used to refer to the territories on the Central American isthmus.

In German, *Zentralamerika* may be used to refer to the territories on the Central American isthmus.

The UN geoscheme defines the region as all states of mainland North America south of the United States; conversely, the European Union excludes Belize and Mexico from its definition of the region.

The Central American Republics

The Federal Republic of Central America, known as the United Provinces of Central America in its first year of creation, was a sovereign state in Central America, which consisted of the territories of the former Captaincy General of Guatemala. A republican democracy, it existed from July 1823 to 1840.

The coat of arms on the nation's flag from 1823–1824 referred to the federation (in Spanish) as *Provincias Unidas del Centro de America* ("United Provinces of the Center of America"); however, its 1824 constitution, coat of arms, and flag called it *Republica Federal de Centroamerica / Centro America* ("Federal Republic of Central America").

It is also sometimes incorrectly referred to in English as the United States of Central America. The flag was introduced to the area by Commodore Louis-Michel Aury inspired by the Argentine flag. The term United Provinces was also used in Argentina's first title "Provincias Unidas del Rio de la Plata" (United Provinces of the Silver River). Commodore Aury established the first independent republic in Old Providence Island (Isla de Providencia) in 1818, off the coast of Nicaragua.

The republic consisted of the states of Guatemala, El Salvador, Honduras, Nicaragua, and Costa Rica (Panama had by then joined Bolivar's Republica de Colombia in 1821). In the 1830s, an additional sixth state was added – Los Altos, with its capital in Quetzaltenango – occupying parts of what are now the western highlands of Guatemala and Chiapas. Although the new nation was now independent of Spain, it had been annexed by Mexico.

The annexation was the focus of disagreement, some seeing the Mexican constitution with its abolition of slavery and establishment of free trade as an improvement over the status quo. During the period of 1838–1840, the federation engaged in civil war by Conservatives fighting against the Liberals. Without a sustained struggle for independence to cement a sense of national identity, the various political factions were unable to overcome their ideological differences and the federation dissolved after a series of bloody conflicts.

Politics

Central American liberals had high hopes for the federal republic, which they believed would evolve into a modern, democratic nation, enriched by trade passing through it between the Atlantic and the Pacific oceans. These aspirations are reflected in the emblems of the federal republic: the flag shows a white band between two blue stripes, representing the land between two oceans. The coat of arms shows five mountains (one for each state) between two oceans, surmounted by a Phrygian cap, the emblem of the French Revolution.

In practice, however, the federation faced insurmountable problems. As a president of Mexico, Vicente Filisola occupied Guatemala City after the formation of Federal Republic of Central America and was successful in annexing El Salvador in 1822, causing an uprising there. In compliance with the Mexican constitution, Filisola convened the Central American congress which forthwith declared its independence from Mexico. Filisola was not able to maintain a fighting force, and his troops were sent back to Mexico by the residents of Guatemala City who paid for their transportation.

The liberal democratic project was strongly opposed by conservative factions allied with the Roman Catholic clergy and the wealthy landowners. Transportation and communication routes between the states were extremely deficient. The bulk of the population lacked any sense of commitment towards the broader federation, perhaps owing to their continued loyalty to the Roman Catholic Church in Spain.

The federal bureaucracy in Guatemala City proved ineffectual, and fears of Guatemalan domination of the union led to protests that resulted in the relocation of the capital to San Salvador in 1831. Wars soon broke out between various factions both in the federation and within individual states. The poverty and extreme political instability of the region prevented the construction of an inter-oceanic canal, from which Central America could have obtained considerable economic benefits.

Dissolution of the Union

The union dissolved in civil war between 1838 and 1840. Its disintegration began when Nicaragua separated from the federation on November 5, 1838, followed by Honduras and Costa Rica. The union effectively dissolved in 1840, by which time four of its five states had declared independence. The union was only officially ended upon El Salvador's self-proclamation of the establishment of an independent republic in February 1841. Because of the chaotic nature of this period an exact date does not exist, but on May 31, 1838, the congress met to declare that the provinces were free to create their own independent republics. In reality, they were just making legal the process of disintegration that had already begun.

Later Attempts at a Federal Union of Central American States

Various attempts were made to reunite Central America in the nineteenth and early twentieth centuries, but none succeeded for any length of time:

The first attempt was in 1842 by former President Francisco Morazán, who became involved in a struggle for control over Costa Rica. After taking control over the capital, Morazán announced he would create a large army to re-create the Federal Republic. Popular feeling rapidly turned against him and a sudden revolt resulted in his arrest and execution by firing squad in September of that year.

A second attempt was made in October 1852 when El Salvador, Honduras and Nicaragua created a Federation of Central America (Federación de Centro America). The union lasted less than a month.

Guatemalan President General Justo Rufino Barrios attempted to reunite the nation by force of arms in the 1880s but he died in battle near the town of Chalchuapa, El Salvador.

A third union of Honduras, Nicaragua, and El Salvador as the Greater Republic of Central America or "Republica Mayor de Centroamerica" lasted from 1896 to 1898.

The latest attempt occurred between June 1921 and January 1922 when El Salvador, Guatemala, Honduras and Costa Rica formed a (second) Federation of Central America. The treaty establishing this federation was signed at San Jose, Costa Rica on January 19, 1921. The treaty stipulated for the future creation of one state of all the four signatories, under one constitution. This second Federation was nearly moribund from the start, having only a Provisional Federal Council made up of delegates from each state.

Despite the failure of a lasting political union, the sense of shared history and the hope for eventual reunification persist in the nations formerly in the union. In 1856–1857 the region successfully established a military coalition to repel an invasion by U.S. adventurer William Walker. Today, all five nations fly flags that retain the old federal motif of two outer blue bands bounding an inner white stripe. (Costa Rica, modified its flag significantly in 1848 by darkening the blue and adding a double-wide inner red band, in honour of the French tricolour.) The short lived sixth state of Los Altos was forcibly annexed by Guatemala's President General Rafael Carrera.

The West Indies

The West Indies Federation, also known as the Federation of the West Indies, was a short-lived Caribbean federation that existed from January 3, 1958, to May 31, 1962. It consisted of several Caribbean colonies of the United Kingdom. The expressed intention of the Federation was to create a political unit that would become independent from Britain as a single state—possibly similar to the Canadian Confederation, Australian Federation, or Central African Federation; however, before that could happen, the Federation collapsed due to internal political conflicts.

Population and Geography

The total population of the West Indies Federation was between 3 and 4 million people, with the majority being of black West African descent. Minorities included Indians from the subcontinent (called East Indians), Europeans, Chinese, and Caribs. There was also a large population of mixed descent (mainly mulattos, but also Afro-Indian, Euro-Indian and mixed-Chinese). In terms of religion, most of the population was Protestant, with significant numbers of Catholics and some Hindus and Muslims (both almost exclusively from the East Indian population).

The West Indies Federation (or just West Indies) consisted of around 24 main inhabited islands and approximately 220–230 minor offshore islands, islets and cays (some inhabited, some uninhabited). The largest island was Jamaica, located in the far northwest of the Federation. To the southeast lay the second largest island, Trinidad, followed by Barbados (in terms of population), located at the eastern extremity of the Federation.

The Federation spanned across all the island groupings in the Caribbean:

The Greater Antilles: Jamaica and the Cayman Islands and the Turks and Caicos Islands

The Lesser Antilles: Barbados, east of the Windward Islands

Leeward Islands: Antigua and Barbuda, Saint Christopher-Nevis-Anguilla, and Montserrat

Windward Islands: Dominica, Saint Lucia, Saint Vincent and the Grenadines, Grenada.

At its widest (west to east), from the Cayman Islands to Barbados it spanned some 2,425 kilometres (1,310 nmi) (and across approximately 22 degrees of longitude) and from the Turks and Caicos Islands in the north, to the Icacos Point, Trinidad in the south it extended 1,700 kilometres (920 nmi) (and across 12 degrees of latitude). However, most of the area along either of these distances was taken up by open water (with the exception of some of the other islands lying in between). By comparison Great Britain stretches across nearly 10 degrees of latitude and Spain extends across almost 20 degrees of longitude. Even though the West Indies was spread across such a vast area, most of its provinces were mostly contiguous and clustered fairly close together in the Eastern Caribbean, with the obvious exceptions of Jamaica, the Cayman Islands, and Turks and Caicos Islands.

Most of the islands have mountainous interiors surrounded by narrow coastal plains. The exceptions were Anguilla, Antigua, Barbuda, the Cayman Islands, the Turks and Caicos Islands (which are all fairly flat), and Trinidad (which has a large mountain range in the north and a small central mountain range in the interior of the otherwise flat island). The narrow coastal plains as well as historical trade is the main reason why almost all of the major settlements (cities and towns) of the Federation were located on the coast. Chief towns included Kingston, Port of Spain, Bridgetown, Spanish Town, Montego Bay, Mandeville, Castries, Roseau, St. George's, Kingstown, St. John's, and Basseterre.

The climate in all the islands is tropical, with hot and humid weather, although inland regions in the larger islands have more temperate climates. Regions falling within the rain shadows (southern coasts of Jamaica and Trinidad and eastern coasts of the Lesser Antilles) are relatively drier. There are two seasons annually: the dry season for the first six months of the year, and the rainy season (also known as the hurricane season) in the second half of the year. Many of the islands fall within the traditional hurricane belt, with the exception of Trinidad (although it occasionally experiences low latitude hurricanes) and thus are at risk from potential wind and flood damage.

Britain classified the Federation as being part of its "Caribbean and North Atlantic Territories" region which was shared alongside other possessions such as Bermuda.

The Federation today is geographically considered to be part of the North American continent as all of its islands are in and around Caribbean, even though Trinidad is located just off-shore from South America and lies on the same continental shelf. *See Bicontinental countries.*

Provinces

The ten provinces or unit territories of the West Indies Federation were:

- Antigua and Barbuda
- Barbados
- Dominica
- Grenada.
- Jamaica (to which were attached the Cayman Islands and Turks and Caicos Islands as dependencies)
- Montserrat
- Saint Christopher-Nevis-Anguilla (present day Saint Kitts and Nevis and Anguilla)
- Saint Lucia
- Saint Vincent and the Grenadines
- Trinidad and Tobago.

Historically "West Indian" nations The Bahamas, Bermuda, Belize, the British Virgin Islands, and Guyana opted not to join because they believed that their future lay with association with North America (for both the Bahamas and Bermuda), Central America, the United States Virgin Islands, and South

America respectively. However, the Bahamas did participate in the 1960 West Indies Federation Games, with a future prime minister of the Bahamas, Perry Christie, as an athlete. There was however possible Interest by Guyana in a very loose re-attempted Caribbean Federation around 1971.

Government and Legal Status

The Federation was an internally self-governing, federal state made up of ten provinces, all British colonial possessions. The federation was created by the United Kingdom in 1958 from most of the British West Indies. Britain intended that the Federation would shortly become a fully independent state, thus simultaneously satisfying the demands for independence from all the colonies in the region. However, the project was doomed by political squabbling among the provinces, and the Federation never achieved full sovereignty, either as a Commonwealth realm or as a republic within the Commonwealth.

The legal basis for the federation was the British Caribbean Federation Act 1956, and the date of formation—January 3, 1958—was set by an Order-in-Council proclaimed in 1957.

As with all British colonies of the period, Queen Elizabeth II was the head of state, and The Crown was vested with the legislative authority for matters concerning executive affairs, defence and the financing of the Federation. Her representative, Patrick Buchan-Hepburn, 1st Baron Hailes, was given the title of Governor-General rather than that of Governor more typical for a British colony. The title may have reflected the federal nature of the state, or indicated the expectations that the Federation would soon become independent. The Governor-General also had the full power by the British Government to veto any laws passed by the Federation.

The Federal Parliament was bicameral, consisting of a nominated Senate and a popularly elected House of Representatives. The Senate consisted of Nineteen members. These members were appointed by the Governor General, after consulting the respective territorial governments. Two members represented each unit (with only one from Montserrat). The House of Representatives had 45 total elected members – Jamaica had 17 seats, Trinidad and Tobago 10 seats, Barbados

5 seats, Montserrat 1 seat, and the remaining islands 2 seats each.

However the government (executive) would be a Council of State, not a Cabinet. It would be presided over by the Governor-General and consist of the Prime Minister and 10 other officials.

There would also be a Federal Supreme Court consisting of a Chief Justice and three (later five) other Justices.

The proposed site for the capital city was Chaguaramas, a few miles west of Port of Spain, Trinidad and Tobago, but the site was part of a United States naval base. In practice, Port of Spain served as the federal capital for the duration of the Federation's existence.

The First Elections

In preparation for the first federal elections, two Federation-wide parties were organised as confederations of local political parties. Both were organised by Jamaican politicians: the West Indies Federal Labour Party by Norman Manley, and the Democratic Labour Party by Alexander Bustamante. In broad terms, the WIFLP consisted of the urban-based parties throughout the Federation, while the DLP consisted of the rural-based parties. A small third party, the Federal Democratic Party was founded in November 1957 by a group of Trinidadians, although it did not win any seats.

The platforms for the two major national parties were similar in many respects. Both advocated maintaining and strengthening ties with the United Kingdom, United States, and Canada (countries with which the islands had strong cultural and economic links); encouraging and expanding tourism; working to bring British Guiana and British Honduras into the Federation and to obtain loans, financial aid, and technical assistance. Despite these similarities, there were differences. The WIFLP had advocated the encouragement of agriculture while the DLP had promised a climate favourable to both private industry and labour, development of human and economic resources. The WIFLP promised to encourage the Bahamas (in addition to British Guiana and British Honduras) to join the Federation, whereas the DLP did not. The WIFLP also campaigned to establish a central bank for the extension of credit resources and advocated a democratic socialist society and full internal self-government for all the unit territories, whilst avoiding the issues of freedom of movement and a customs union. The DLP said nothing about full internal self-government, attacked socialism, wished to avoid high taxation (via loans and technical aid) and emphasized West Indian unity, freedom of worship and speech, and encouragement of trade unions.

Federal elections were held on March 25, 1958. The WIFLP won the election, winning 26 seats while the DLP carried 19 seats. The bulk of the WIFLP seats came from the smaller islands while the DLP carried the majority in Jamaica and Trinidad & Tobago. The DLP won 11 of the Jamaican seats and 6 of the Trinidadian seats. In appointing the Senate, Governor General Lord Hailes realized that only the St Vincent island government was DLP controlled and as a result the Senate was going to be disproportionately pro WIFLP. In a controversial decision, he contacted the opposition DLP groups in Jamaica and Trinidad, and appointed one DLP senator from each of those islands. Thus the Senate consisted of a total of 15 WIFLP members and 4 DLP members.

WIFLP leader Sir Grantley Adams of Barbados became Prime Minister. The selection of Adams as the Prime Minister was indicative of the problems the Federation would face. The expected leader of the WIFLP was Norman Manley, Premier of Jamaica, and the next logical choice was Dr. Eric Williams, Premier of Trinidad and Tobago. However, neither had contested the Federal elections, preferring to remain in control of their respective island power bases. This suggested that the leaders of the two most important provinces did not see the Federation as viable. Similarly, Alexander Bustamante, the Jamaican founder of the DLP, also declined to contest the Federal election, leaving the party leadership to the Trinidadian Ashford Sinanan. The absence of the leading Jamaican politicians from any role at the federal level was to undermine the Federation's unity.

Government Services

The Federation also had a number of units deemed to be common services for the entire federation, these were :

- The Federal Shipping Service
- Te Federal Supreme Court

- The University College of the West Indies
- The West Indies Meteorological Service
- The West India Regiment.

The Advisory Services of the Federation included:

- Agriculture
- Civil Aviation
- Education
- Fisheries
- Forestry
- Livestock
- Maritime Services
- Marketing
- Medicine
- Postal Services
- Telecommunications.

Federal Problems

The politics of the embryonic Federation were wracked by struggles between the federal government and the provincial governments, and between the two largest provinces (Jamaica and Trinidad and Tobago) and the smaller provinces.

The West Indies Federation had an unusually weak federal structure. For instance, its provinces were not contained in a single customs union. Thus, each province functioned as a separate economy, complete with tariffs, largely because the smaller provinces were afraid of being overwhelmed by the large islands' economies. Also, complete freedom of movement within the Federation was not implemented, as the larger provinces were worried about mass migration from the smaller islands. In this sense, the current European Union can be said to have implemented a more unified economic space than the West Indian attempt.

Nor could the federal government take its component states to task. The initial federal budget was quite small, limiting the federal government's ability to use its financial largesse as a carrot. It was dependent upon grants from the United Kingdom and from its member states. The provincial budgets of Jamaica and Trinidad and Tobago were both larger than the federal budget. This led to repeated requests for those states to provide greater financing to the federal government. These requests were not well received, as Jamaica and Trinidad and Tobago together already contributed 85 percent of the federal revenue, in roughly equal portions.

Furthermore, the office of the Prime Minister was a weak one. Unlike other Westminster systems with Prime Ministers, the West Indian Federation's PM could not dissolve Parliament.

Relationship with Canada

The Federation maintained a particularly close relationship with Canada, which had a similar past in that it was a confederation of several former British colonies. In the early years, several Caribbean leaders suggested that the West Indies Federation should investigate the possibility of becoming a Canadian province, though this was never more than a fleeting interest.

Despite the break down in talks, in May 1961, Canada presented the West Indies Federation with two of the region's most important gifts: two merchant ships, named *The Federal Palm* and *The Federal Maple*. These two vessels visited every island in the federation twice monthly, providing a crucial sea-link between the islands.

Dissolution

Many reasons have been put forward to explain the demise of the federation, some of them detailed in "Problems" above. These include the utter lack of local popular support, competing insular nationalism, the weakness of the federal government, prohibitions on federal taxation and freedom of movement, inadequacies in the Federal constitution, fundamental changes made to the constitution very early in its existence, political feuds between the influential leaders, the decision of the three most influential politicians not to contest Federal elections, friction between these leaders and the Federal government, the overwhelming concentration of population and resources in the two largest units, geographic and cultural distance between the units, the lack of a history of common administration, and the impact of the period of self-government that followed the promotion from Crown Colony system.

However, the immediate catalyst for the dissolution of the Federation was Jamaican discontent. By 1961, there were a number of reasons for Jamaica's dissatisfaction with the state of affairs:

- Jamaica was fairly remote from most of the other islands in the Federation, lying several hundred miles to the west.
- Jamaica's share of the seats in the federal parliament was smaller than its share of the total population of the Federation.
- It was believed that the smaller islands were draining Jamaica's wealth.
- Many in Jamaica were upset that Kingston had not been chosen as the federal capital.

The most important reason for Jamaican dissatisfaction was the Federation's continuing colonial status. Jamaica had joined the Federation because its leaders had believed that the West Indies would quickly be granted independence. Nearly three years after the formation of the Federation, this had not occurred; meanwhile, smaller British colonies, like Cyprus and Sierra Leone, had gained independence. Thus, many Jamaicans believed that the island could and should seek independence in its own right.

There were also problems with the Federation's proposed capital in Chaguaramas, at that time still in the hands of the United States (having leased it as a naval base from the United Kingdom during World War 2). Many of the Caribbean provincial leaders wanted Chaguaramas to be the Federation's capital.

Provincial leaders such as Norman Manley of Jamaica and Dr. Eric Williams pushed for handing over of Chaguramas to the Federation from the US. However the US and the UK disagreed and the Federation's Prime Minister Grantley Adams denied the provincial leaders from obtaining Chaguramas. For many Jamaicans it appeared that the Federation would then just hamper their development and movement towards independence.

As a result the Bustamante-led Jamaica Labour Party (the local component of the West Indian DLP) successfully forced

Manley to hold a referendum in September 1961 on political secession from the Federation. It passed, with 54 percent of the vote, despite the opposition of Manley, the province's Chief Minister at the time. Manley himself lost the subsequent island elections in April 1962, and Bustamante became the first Prime Minister of an independent Jamaica on August 6, 1962.

After Jamaica left, there was an attempt to salvage a new federation from the wreckage of the old. Much depended on Premier Williams of Trinidad and Tobago, who had stated previously that he wanted a "strong federation." Premier Vere Bird of Antigua responded that his province would only be in a federation with Trinidad as an equal partner, not as "a little Tobago." He did indicate that a strong federation was acceptable provided that no attempt was made to create a unitary state.

Negotiations on this new federation began in September 1961; however, they indicated that Trinidad would have to provide 75 to 80 percent of the new Federation's revenue. Also, even though Trinidad would now represent 60 percent of the new Federation's population, the proposals under consideration would give it less than half of the seats in parliament.

By November, Williams indicated that he was now in favour of the idea of a unitary state. Failing that, he resolved to take Trinidad and Tobago into independence. In this, he was buoyed by his re-election as Trinidadian leader on December 4, 1961. Later that December, Premier Errol Barrow of Barbados met with Williams, but failed to persuade him to keep Trinidad in the Federation.

On January 14, 1962, the People's National Movement (the Williams-led Trinidad component of the WIFLP) passed a resolution rejecting any further involvement with the Federation. Williams himself stated that "one from ten leaves nought"—in other words, without Jamaica, no Federation was possible. Trinidad and Tobago became independent on August 31, 1962.

Without Trinidad and Jamaica, the remaining "Little Eight" attempted to salvage some form of a West Indian Federation, this time centred on Barbados. However, these negotiations ultimately proved fruitless. Without its two largest states, the

Federation was doomed to financial insolvency. Barbados now refused to shoulder the financial burden, and Antigua and Grenada began toying with the idea of merging with Jamaica and Trinidad, respectively.

The West Indies Federation was legally dissolved with the Parliament of the United Kingdom's West Indies Act 1962. The remaining "Little Eight" provinces once again became separate colonies supervised directly from London, most of which became independent later on, as follows:

- Barbados – 1966
- Grenada – 1974
- Dominica – 1978
- Saint Lucia – 1979
- St Vincent and the Grenadines – 1979
- Antigua and Barbuda – 1981
- Saint Kitts and Nevis – 1983.

Montserrat remains an overseas territory of the United Kingdom. The Cayman Islands and Turks and Caicos Islands had been separated from Jamaica upon the latter's independence in 1962; Anguilla was separated from Saint Kitts and Nevis in 1980. All three remain UK territories as well.

Legacy

The federation's currency was the West Indies dollar (though Jamaica continued to use the pound), which was later succeeded by the East Caribbean dollar, the Barbadian dollar, and the Trinidad and Tobago dollar. Successor organisations included the West Indies Associated States and CARICOM.

Some see the West Indies cricket team as a legacy of the Federation, although the side was actually organised thirty years prior to the birth of the federation.

Another lasting regional fixture, officially created before the Federation, is the University of the West Indies. During the Federation, the University pursued a policy of regional expansion beyond the main Jamaica campus. Two other campuses were established: one in Trinidad and Tobago, established in 1960, and one in Barbados, established a short time after the Federation dissolved in 1963.

Stamps

During the Federation's existence, each member continued to issue its own postage stamps as before; but on April 22, 1958, each of the members (except for the Cayman Islands) issued a set of three commemorative stamps. All of these stamps used a common design depicting a map of the Caribbean and a portrait of Queen Elizabeth, with an inscription at the top reading "THE WEST INDIES/FEDERATION 1958" at the top and the name of the member at the bottom. All of these stamps are quite common in both mint and used condition.

Prior Attempts at Federation

The Federation of the West Indies was not the first attempt at a British Caribbean federation (nor would it be the last). The history of the previous attempts at federations and unions, in part, explains the failure of the 1958 Federation.

The initial federal attempts never went so far as to try to encompass all of the British West Indies (BWI), but were more regional in scope. The historical regional groupings included the British Leeward Islands, British Windward Islands and Jamaica with nearby colonies..

9

General Introduction to the Northern Andes and Guiana

The Andes are the world's longest continental mountain range. It is a continuous chain of highland along the western coast of South America. The range is over 7,000 km (4,300 mi) long, 200 km (120 mi) to 700 km (430 mi) wide (widest between 18° to 20°S latitude), and of an average height of about 4,000 m (13,000 ft).

Over its length the Andean range is at several locations split into several ranges, often two great ranges, named Cordillera Oriental and the Cordillera Occidental, often separated by an intermediate depression. The Andes mountains extend over seven countries: Argentina, Bolivia, Chile, Colombia, Ecuador, Peru, and Venezuela, some of which are known as Andean States.

The Andes mountain range is the highest mountain range outside Asia. The highest peak, Aconcagua, rises to 6,962 m (22,841 ft) above sea level. The summit of Mount Chimborazo in the Ecuadorean Andes is the point on the Earth's surface most distant from its center, because of the equatorial bulge.

Name

There are many theories about the etymology of the name *Andes*. Some believe *Andes* derives from the Quechua word *anti*, which means "high crest". Some believe *Andes* derives from one of the four regions of the Inca empire, or *Anti(s)*. Some believe *Andes* derives from the Spanish word *anden* which means terrace in reference to the cultivation terraces used by the Incas and other related peoples. Some believe *Andes* is Spanish shorthand for "Andenes" or "Andenerias".

Geography

The Andes can be divided into three sections: the Southern Andes in Argentina and Chile; the Central Andes, including the Chilean and Peruvian cordilleras and parts of Bolivia; and the northern section in Venezuela, Colombia, and Ecuador consisting of two parallel ranges, the Cordillera Occidental and the Cordillera Oriental. In Colombia, north to the border with Ecuador, the Andes split in three parallel ranges, western, central and eastern. (cordillera occidental, central y oriental). In the northern part the isolated Sierra Nevada de Santa Marta range is often considered part of the Andes. The eastern range is the only one that extends to Venezuela. The term *cordillera* comes from the Spanish word meaning 'rope'. The Andes range is approximately 200 km (124 mi) wide throughout its length, except in the Bolivian flexure where it is 640 km (398 mi) wide. The islands of the Dutch Caribbean Aruba, Bonaire, and Curacao, which lie in the Caribbean Sea off the coast of Venezuela, represent the submerged peaks of the extreme northern edge of the Andes range.

Geology

The Andes is a Mesozoic-Tertiary orogenic belt of mountains along the Pacific Ring of Fire, a zone of volcanic activity and orogeny that encompasses the Pacific rim of the Americas as well as the Asia-Pacific region. The Andes are the result of plate tectonics processes, caused by the subduction of oceanic crust beneath the South American plate. The main cause of the rise of the Andes is the compression of western rim of the South American Plate due to the subduction of the Nazca Plate and the Antarctic Plate. To the east the Andes is bounded by several sedimentary basins such as Orinoco, Amazon Basin, Madre de Dios and Gran Chaco which separates the Andes from the ancient cratons in eastern South America. In the south the Andes shares a long boundary with the former Patagonia Terrane. To the west the Andes ends at the Pacific Ocean, although the Peru-Chile trench can be considerated its ultimate western limit. From a geographical approach the Andes is considered to have its western boundaries marked by the apparition of coastal low lands and a less abrupt topography.

Orogeny

The western rim of the South American Plate has been the place of several pre-Andean orogenies since at least the Pampean orogeny of late Proterozoic. These early orogenic events led to amalgamation of several terranes and microcontinents onto the South American part of Gondwana.

The formation of the modern Andes began in the Jurassic Period. It was during the Cretaceous Period that the Andes began to take their present form, by the uplifting, faulting and folding of sedimentary and metamorphic rocks of the ancient cratons to the east. Tectonic forces along the subduction zone along the entire west coast of South America where the Nazca Plate and a part of the Antarctic Plate are sliding beneath the South American Plate continue to produce an ongoing orogenic event resulting in minor to major earthquakes and volcanic eruptions to this day. In the extreme south a major transform fault separates Tierra del Fuego from the small Scotia Plate. Across the 1,000 km (620 mi) wide Drake Passage lie the mountains of the Antarctic Peninsula south of the Scotia Plate which appear to be a continuation of the Andes chain.

Volcanism

Astronaut photograph with the high plains of the Andes Mountains in the foreground, with a line of young volcanoes facing the much lower Atacama Desert.

The Andes range has many active volcanoes, which are distributed in four volcanic zones separated by areas of inactivity. The Andean volcanism is a result of subduction of the Nazca Plate and Antarctic Plate underneath the South American Plate. The belt is subdivided into four main volcanic zones that are separed from each other by volcanic gaps. The volcanoes of the belt are diverse in terms of activity style, products and morphology. While some differences can be explained by which volcanic zone a volcano belongs to, there significant differences inside volcanic zones and even between neighboring volcanoes. Despite being a type location for calc-alkalic and subduction volcanism, the Andean Volcanic Belt has a large range of volcano-tectonic settings, such as rift systems and extensional zones, transpressional faults, subduction of mid-ocean ridges and seamount chains apart from a large range on crustal thicknesses and magma ascent paths, and different amount of crustal assimilations.

Climate

The climate in the Andes varies greatly depending on location, altitude, and proximity to the sea. The southern section is rainy and cool, the central Andes are dry. The northern Andes are typically rainy and warm, with an average temperature of 18 °C (64 °F) in Colombia. The climate is known to change drastically in rather short distances. Rainforests exist just miles away from the snow covered peak Cotopaxi. The mountains have a large effect on the temperatures of nearby areas. The snow line depends on the location. It is at between 4,500–4,800 m (14,800–15,800 ft) in the tropical Ecuadorian, Colombian, Venezuelan, and northern Peruvian Andes, rising to 4,800–5,200 m (15,800–17,060 ft) in the drier mountains of southern Peru south to northern Chile south to about 30°S, then descending to 4,500 m (14,760 ft) on Aconcagua at 32°S, 2,000 m (6,600 ft) at 40°S, 500 m (1,640 ft) at 50°S, and only 300 m (980 ft) in Tierra del Fuego at 55°S; from 50°S, several of the larger glaciers descend to sea level.

The Andes of Chile and Argentina can be divided in two climatic and glaciological zones; the Dry Andes and the Wet Andes. Since the Dry Andes extends from the latitudes of Atacama Desert to the area of Maule River, precipitation is more sporadical and there are strong temperature oscillations. The line of equilibrium may shift drastically over short periods of time, leaving a whole glacier in the ablation area or in the accumulation area.

Flora

Rainforests used to encircle much of the northern Andes but are now greatly diminished, especially in the Choco and inter-Andean valleys of Colombia. As a direct opposite of the humid Andean slopes are the relatively dry Andean slopes in most of western Peru, Chile and Argentina. Along with several Interandean Valles, they are typically dominated by deciduous woodland, shrub and xeric vegetation, reaching the extreme in the slopes near the virtually lifeless Atacama Desert.

About 30,000 species of vascular plants live in the Andes with roughly half being endemic to the region, surpassing the diversity of any other hotspot.

The small tree *Cinchona pubescens*, a source of quinine which is used to treat malaria, is found widely in the Andes as far south as Bolivia. Other important crops that originated from the Andes are tobacco and potatoes. The high-altitude *Polylepis* forests and woodlands are found in the Andean areas of Colombia, Ecuador, Peru, Bolivia and Chile.

These trees, by locals referred to as Quenua, Yagual and other names, can be found at altitudes of 4,500 m (14,760 ft) above sea level.

It remains unclear if the patchy distribution of these forests and woodlands is natural, or the result of clearing which began during the Incan period. Regardless, in modern times the clearance has accelerated, and the trees are now considered to be highly endangered, with some believing that as little as 10% of the original woodland remains.

Fauna

The Andes are rich in fauna: With almost 1,000 species, of which roughly 2/3 are endemic to the region, the Andes is the most important region in the world for amphibians. Animal diversity in the Andes is high, with almost 600 species of mammals (13% endemic), more than 1,700 species of birds (c. 1/3 endemic), more than 600 species of reptiles (c. 45% endemic), and almost 400 species of fish (c. 1/3 endemic).

The Vicuna and Guanaco can be found living in the Altiplano, while the closely related domesticated Llama and Alpaca are widely kept by locals as pack animals and for their meat and wool. The nocturnal chinchillas, two threatened members of the rodent order, inhabits the Andes' alpine regions.

The Andean Condor, the largest bird of its kind in the Western Hemisphere, occurs throughout much of the Andes but generally in very low densities. Other animals found in the relatively open habitats of the high Andes include the huemul, cougar, foxes in the genus *Pseudalopex*, and, for birds, certain species of tinamous (notably members of the genus *Nothoprocta*), Andean Goose, Giant Coot, flamingos (mainly associated with hypersaline lakes), Lesser Rhea, Andean Flicker, Diademed Sandpiper-plover, miners, sierra-finches and diucafinches.

Lake Titicaca hosts several endemics, among them the highly endangered Titicaca Flightless Grebe and Titicaca Water Frog. A few species of hummingbirds, notably some hillstars, can be seen at altitudes above 4,000 m (13,100 ft), but far higher diversities can be found at lower altitudes, especially in the humid Andean forests ("cloud forests") growing on slopes in Colombia, Ecuador, Peru, Bolivia and far northwestern Argentina.

These forest-types, which includes the Yungas and parts of the Choco, are very rich in flora and fauna, although few large mammals exists, exceptions being the threatened Mountain Tapir, Spectacled Bear and Yellow-tailed Woolly Monkey.

Birds of humid Andean forests include mountain-toucans, quetzals and the Andean Cock-of-the-rock, while mixed species flocks dominated by tanagers and Furnariids commonly are seen-in contrast to several vocal but typically cryptic species of wrens, tapaculos and antpittas.

A number of species such as the Royal Cinclodes and White-browed Tit-spinetail are associated with *Polylepis*, and consequently also threatened.

Human Activity

The Andes mountains form a north-south axis of cultural influences. A long series of cultural development culminated in the expansion of the Inca civilization and Inca Empire in the central Andes during the 15th century. The Incas formed this civilization through imperialistic militarism as well as careful and meticulous governmental management. The government sponsored the construction of aqueducts and roads in addition to preexisting installations. Some of these constructions, as those created by the Roman Empire, are still in existence today.

Devastated by European diseases to which they had no immunity, and civil wars, in 1532 the Incas were defeated by an alliance composed of tens of thousands allies from nations they had subjugated (e.g. Huancas, Chachapoyas, Canaris) and a small army of 180 Spaniards led by Francisco Pizarro.

One of the few Inca sites the Spanish never found in their conquest was Machu Picchu, which lay hidden on a peak on the eastern edge of the Andes where they descend to the Amazon. The main surviving languages of the Andean peoples are those of the Quechua and Aymara language families. Woodbine Parish and Joseph Barclay Pentland surveyed a large part of the Bolivian Andes from 1826 to 1827.

In modern times, the largest Andean cities are Bogota in Colombia with a population of 8 million and Santiago in Chile. The largest city of the Andean zone is the Peruvian capital, Lima, which is located a few metres above sea level.

Transportation

Several major cities exist in the Andes, among them Bogota and Cali in Colombia, Quito in Ecuador, Merida in Venezuela, La Paz in Bolivia, and Arequipa and Cusco in Peru.

These and most other cities are now connected with asphalted roads, while smaller towns are often connected by dirt roads, which may require a 4x4 vehicle. Because of the arduous terrain, localities where vehicles are of little use remain. Locally, Llamas continue to play an important role as pack animals, but this use has generally diminished in modern times.

Agriculture

The ancient peoples of the Andes such as the Incas have practiced irrigation techniques for over 6,000 years. Because of the mountain slopes, terracing has been a common practice. Terracing, however, was only extensively employed after Incan imperial expansions to fuel their expanding realm.

The potato holds a very important role as an internally consumed staple crop. Maize was also an important crop for these people. However, they were mainly used for the production of the culturally important chicha. Currently, tobacco, cotton and coffee are the main export crops. Coca, despite eradication programmes in some countries, remains an important crop for legal local use in a mildly stimulating herbal tea, and, both controversially and illegally, for the production of cocaine.

Mining

The Andes rose to fame for its mineral wealth during the Spanish conquest of South America. Although Andean Amerindian peoples crafted ceremonial jewelry of gold and other metals the mineralizations of the Andes were first mined in large scale after the Spanish arrival.

Potosi in present-day Bolivia was one of the principal mines of the Spanish Empire in the New World. Rio de la Plata and Argentina derive their names from the silver of Potosi.

Currently, mining in the Andes of Chile and Peru place these countries as the 1st and 3rd major producers of copper in the world. The Bolivian Andes produce principally tin although historically silver mining had a huge impact on the economy of 17th century Europe.

There is a long history of mining in the Andes, from the Spanish silver mines in Potosi in the 16th century to the vast current porphyry copper deposits of Chuquicamata and Escondida in Chile and Toquepala in Peru. Other metals including iron, gold and tin in addition to non-metallic resources are also important.

Guiana Space Centre

The Guiana Space Centre or, more commonly, Centre Spatial Guyanais (CSG) is a French spaceport near Kourou in French Guiana. Operational since 1968, it is particularly suitable as a location for a spaceport due to its proximity to the equator, and that launches are in a favourable direction over water. The European Space Agency, the French space agency CNES, and the commercial Arianespace company conduct launches from Kourou.

The location was selected in 1964 to become the spaceport of France. When the European Space Agency (ESA) was founded in 1975, France offered to share Kourou with ESA. Commercial launches are bought also by non-European companies. ESA pays two thirds of the spaceport's annual budget, and has also financed the upgrades made during the development of the Ariane launchers.

The Guianas

The Guyanas or The Guianas (in Spanish: Guayanas) refers to a region in north-eastern South America which includes the following three territories:

- Guyana, formerly British Guiana
- Suriname, formerly Dutch Guiana
- French Guiana, an overseas department of France.

Historically the following adjacent areas were also included in The Guianas:

- Portuguese Guiana (Brazilian Guiana), in north-west Brazil, originally part of the Portuguese Empire
- Spanish Guiana (Guayana Region), in south-east Venezuela, originally part of the Spanish Empire.

10

Guyana

Guyana, officially the Co-operative Republic of Guyana and previously known as British Guiana, is a state on the northern coast of South America that is culturally part of the Anglophone Caribbean. Guyana has been a former colony of the Spanish, French, Dutch, and British. It is the only state of the Commonwealth of Nations on mainland South America, and is also a member of the Caribbean Community (CARICOM), which has its secretariat headquarters in Guyana's capital; Georgetown. Guyana achieved independence from the United Kingdom in 26 May 1966 and became a Republic on 23 February 1970; it is also one of the youngest countries.

Historically, the region known as "Guiana" or "Guayana" was the large shield landmass north of the Amazon River and East of the Orinoco River. Five sub-regions were carved out of the landmass by colonial powers in the late 17th and early 18th century: Spanish Guiana, Portuguese Guiana, British Guiana, Dutch Guiana, and the present French overseas department of French Guiana. Historic Guyana is made up of three Dutch colonies; Essequibo, Demerara and Berbice while modern Guyana is bordered to the east by Suriname, to the south and southwest by Brazil, to the west by Venezuela, and on the north by the Atlantic Ocean. At 215,000 km^2, Guyana is the third-smallest independent state on the mainland of South America (after Uruguay and Suriname). Its population is approximately 770,000. It is one of the four non-Spanish-speaking territories on the continent, along with the countries of Brazil (Portuguese), Suriname (Dutch), and French Guiana (French).

Etymology

The name "Guyana" is derived from *guiana* the original name for the region which now includes Guyana, Suriname, French Guiana and parts of Venezuela and Brazil. It is believed the original term came from a Taino word meaning either "land of fast-flowing water" (a reference to the numerous rivers of the territory) or "respectable", but the most famous and permanent one is "The Land Of Many Waters".

History

Guyana was inhabited by the Arawak and Carib tribes of Native Americans. Although Christopher Columbus sighted Guyana during his third voyage (in 1498), the Dutch were the first to establish colonies: Essequibo (1616), Berbice (1627), and Demerara (1752). The British assumed control in the late 18th century, and the Dutch formally ceded the area in 1814. In 1831 the three separate colonies became a single Britsh colony known as British Guiana.

The first slaves were forcibly brought to Guyana under harsh conditions in the early 17th century in order to provide free labour for the development of the local economy and to import goods to Europe. Escaped slaves formed their own settlements known as Maroon communities. With the abolition of slavery in 1834 and after the five year apprenticeship period, many of the former slaves bought and paid for land which they then used to form other communities e.g. Buxton, Plaissance, Victoria, Queenstown, etc. Indentured labourers from modern-day Portugal (1834), Germany (first in 1835), Ireland (1836), Scotland (1837), Malta (1839), China and eastern India (Madras, Bengal and Bihar primarily, beginning in 1838) were imported to work on the sugar plantations.

Since Independence in 1824, Venezuela has claimed the western land to the Essequibo river. Letters from Simon Bolivar warned the British Kingdom about the Berbice and Demerara settlers settling on land the Venezuelans claimed was theirs. In 1899, an international tribunal, with no Venezuelan representation, ruled the land belonged to United Kingdom.

Guyana achieved independence from the United Kingdom on 26 May 1966 and became a republic on 23 February 1970, remaining a member of the Commonwealth. The United States

State Department and the US Central Intelligence Agency (CIA), along with the British government, played a strong role in influencing political control in Guyana during this time. The American government supported Forbes Burnham during the early years of independence because Cheddi Jagan was a self declared Marxist. They provided secret financial support and political campaign advice to Burnham's People's National Congress to the detriment of the Jagan-led People's Progressive Party, mostly supported by Guyanese of Indian descent.

In 1978, Guyana received considerable international attention when 918 almost entirely American members (more than 300 of whom were children) of the Jim Jones-led Peoples Temple died in a mass murder/suicide in Jonestown – a settlement created by the Peoples Temple. An attack by Jim Jones' body guards at a small remote airstrip close to Jonestown resulted in the murder of five people, including Leo Ryan, the only congressman ever murdered in the line of duty in US history.

Geography

The Geography of Guyana comprises the physical characteristics of the country in Northern South America and part of Caribbean South America, bordering the North Atlantic Ocean, between Suriname and Venezuela, with a land area of approximately 214,970 square kilometres. The country is situated between 1 and 9 north latitude and between 56 and 62 west longitude. With a 430-kilometre Atlantic coastline on the northeast, Guyana is bounded by Venezuela on the west, Brazil on the west and south, and Suriname on the east. The land comprises three main geographical zones: the coastal plain, the white sand belt, and the interior highlands.

Description

The coastal plain, which occupies about 5% of the country's area, is home to more than 90% of its inhabitants. The plain ranges from five to six kilometers wide and extends from the Courantyne River in the east to the Venezuelan border in the northwest.

The coastal plain is made up largely of alluvial mud swept out to sea by the Amazon River, carried north by ocean currents, and deposited on the Guyanese shores. A rich clay of great fertility, this mud overlays the white sands and clays formed from the erosion of the interior bedrock and carried seaward by the rivers of Guyana. Because much of the coastal plain floods at high tide, efforts to dam and drain this area have gone on since the 1700s.

Guyana has no well-defined shoreline or sandy beaches. Approaching the ocean, the land gradually loses elevation until it includes many areas of marsh and swamp. Seaward from the vegetation line is a region of mud flats, shallow brown water, and sandbars. Off New Amsterdam, these mud flats extend almost twenty-five kilometers. The sandbars and shallow water are a major impediment to shipping, and incoming vessels must partially unload their cargoes offshore in order to reach the docks at Georgetown and New Amsterdam.

A line of swamps forms a barrier between the white sandy hills of the interior and the coastal plain. These swamps, formed when water was prevented from flowing onto coastal croplands by a series of dams, serve as reservoirs during periods of drought.

The white sand belt lies south of the coastal zone. This area is 150 to 250 kilometers wide and consists of low sandy hills interspersed with rocky outcroppings. The white sands support a dense hardwood forest. These sands cannot support crops, and if the trees are removed erosion is rapid and severe. Most of Guyana's reserves of bauxite, gold, and diamonds are found in this region.

The largest of Guyana's three geographical regions is the interior highlands, a series of plateaus, flat-topped mountains, and savannahs that extend from the white sand belt to the country's southern borders. The Pacaraima Mountains dominate the western part of the interior highlands. In this region are found some of the oldest sedimentary rocks in the Western Hemisphere. Mount Roraima, on the Venezuelan border, is part of the Pakaraima range and, at 2,762 meters, is Guyana's tallest peak. Farther south lies the Kaieteur Plateau, a broad, rocky area about 600 meters in elevation; the 1,000-meter high Kanuku Mountains; and the low Acarai Mountains situated on the southern border with Brazil.

Much of the interior highlands consist of grassland. The largest expanse of grassland, the Rupununi Savannah, covers about 15,000 square kilometers in southern Guyana. This savannah also extends far into Venezuela and Brazil. The part in Guyana is split into northern and southern regions by the Kanuku Mountains.

The sparse grasses of the savannah in general support only grazing, although Amerindian groups cultivate a few areas along the Rupununi River and in the foothills of the Kanuku Mountains.

Hydrology

Guyana is a water-rich country. Numerous rivers flow into the Atlantic Ocean, generally in a northward direction. A number of rivers in the western part of the country, however, flow eastward into the Essequibo River, draining the Kaieteur Plateau. The Essequibo, the country's major river, runs from the Brazilian border in the south to a wide delta west of Georgetown. The rivers of eastern Guyana cut across the coastal zone, making east-west travel difficult, but they also provide limited water access to the interior. Waterfalls generally limit water transport to the lower reaches of each river. Some of the waterfalls are spectacular; for example, Kaieteur Falls on the Potaro River drops 227 metres, more than four times the height of Niagara Falls.

Drainage throughout most of Guyana is poor and river flow sluggish because the average gradient of the main rivers is only one meter every five kilometers. Swamps and areas of periodic flooding are found in all but the mountainous regions, and all new land projects require extensive drainage networks before they are suitable for agricultural use. The average square kilometre on a sugar plantation, for example, has six kilometers of irrigation canals, eighteen kilometers of large drains, and eighteen kilometers of small drains.

These canals occupy nearly one-eighth of the surface area of the average sugarcane field. Some of the larger estates have more than 550 kilometers of canals; Guyana itself has a total of more than 8,000 kilometers. Even Georgetown is below sea level and must depend on dikes for protection from the Demerara River and the Atlantic Ocean.

Climate

Lying near the equator, Guyana has a tropical climate, and temperatures do not vary much throughout the year. The year has two wet seasons, from December to early February and from late April to mid-August.

Although the temperature never gets dangerously high, the combination of heat and humidity can at times seem oppressive. The entire area is under the influence of the northeast trade winds, and during the midday and afternoon sea breezes bring relief to the coast. Guyana lies south of the path of Caribbean hurricanes and none is known to have hit the country.

Temperatures in Georgetown are quite constant, with an average high of 32 °C (89.6 °F) and an average low of 24 °C (75.2 °F) in the hottest month (July), and an average range of 29 to 23 °C (84.2 to 73.4 °F) in February, the coolest month. The highest temperature ever recorded in the capital was 37.2 °C (99.0 °F) and the lowest 16.6 °C (61.9 °F). Humidity averages 70 percent year-round. Locations in the interior, away from the moderating influence of the ocean, experience slightly wider variations in daily temperature, and nighttime readings as low as 12 °C (53.6 °F) have been recorded. Humidity in the interior is also slightly lower, averaging around 60 percent.

Rainfall is heaviest in the northwest and lightest in the southeast and interior. Annual averages on the coast near the Venezuelan border are near 2,500 millimetres (98.4 in), farther east at New Amsterdam, between 2,000 and 1,500 millimetres (78.7 and 59.1 in) in southern Guyana's Rupununi Savannah. Areas on the northeast sides of mountains that catch the trade winds average as much as 3,500 millimetres (137.8 in) of precipitation annually. Although rain falls throughout the year, about 50 percent of the annual total arrives in the summer rainy season that extends from May to the end of July along the coast and from April through September farther inland. Coastal areas have a second rainy season from November through January. Rain generally falls in heavy afternoon showers or thunderstorms. Overcast days are rare; most days include four to eight hours of sunshine from morning through early afternoon.

Characteristics

Area

Total: 214,970 km²

Land: 196,850 km²

Water: 18,120 km²

Land Boundaries

Total: 2,462 km

Border countries: Brazil 1,119 km, Suriname 600 km, Venezuela 743 km

Coastline

459 km (285 mi)

Maritime Claims

Continental shelf: 200 nmi (370.4 km; 230.2 mi) or to the outer edge of the continental margin

Exclusive fishing zone: 200 nmi (370.4 km; 230.2 mi)

Territorial sea: 12 nmi (22.2 km; 13.8 mi)

Terrain

Mostly rolling highlands; low coastal plain; savanna in south.

Elevation Extremes

Lowest point: Atlantic Ocean, 0 m

Highest point: Mount Roraima, 2,835 m

Natural Resources

Bauxite, gold, diamonds, hardwood timber, shrimp, fish.

Land Use

Arable land:	2%
Permanent crops:	0%
Permanent pastures:	6%
Forests and woodland:	84%
Other:	8% (1993 est.).

Irrigated Land

1,300 km² (1993 est.)

Natural Hazards

Flash floods are a constant threat during rainy seasons

Environment, Current Issues

Water pollution from sewage and agricultural and industrial chemicals; deforestation

Guyana can be divided into five natural regions; a narrow and fertile marshy plain along the Atlantic coast (low coastal plain) where most of the population lives; a white sand belt more inland (hilly sand and clay region), containing most of Guyana's mineral deposits; the dense rain forests (Forested Highland Region) in the middle of the country; the grassy savannah in the southern west; and the larger interior highlands (interior savannah) consisting mostly of mountains that gradually rise to the Brazilian border.

Some of Guyana's highest mountains are Mount Ayanganna (2,042 metres /6,699 feet), Monte Caburai (1,465 metres /4,806 feet) and Mount Roraima (2,835 metres /9,301 feet — the highest mountain in Guyana) on the Brazil-Guyana-Venezuela tripoint border, part of the Pakaraima range. Mount Roraima and Guyana's table-top mountains (tepuis) are said to have been the inspiration for Sir Arthur Conan Doyle's 1912 novel *The Lost World*. There are also many steep escarpments and waterfalls, including Kaieteur Falls. Between the Rupununi River and the border with Brazil lies the Rupununi savannah, south of which lie the Kanuku Mountains.

The four longest rivers are the Essequibo at 1,010 kilometres (628 mi) long, the Courantyne River at 724 kilometres (450 mi), the Berbice at 595 kilometres (370 mi), and the Demerara at 346 kilometres (215 mi). The Courantyne river forms the border with Suriname. At the mouth of the Essequibo are several large islands, including the 145 km (90 mi) wide Shell Beach lies along the northwest coast, which is also a major breeding area for sea turtles (mainly Leatherbacks) and other wildlife.

The local climate is tropical and generally hot and humid, though moderated by northeast trade winds along the coast.

There are two rainy seasons, the first from May to mid-August, the second from mid-November to mid-January.

Guyana has one of the largest unspoiled rainforests in South America, some parts of which are almost inaccessible by humans. The rich natural history of Guyana was described by early explorers Sir Walter Raleigh and Charles Waterton and later by naturalists Sir David Attenborough and Gerald Durrell. In 2008, the BBC ran a three-part programme called *Lost Land of the Jaguar* which highlighted the huge diversity of wildlife, including undiscovered species and rare species such as the giant otter and harpy eagle.

Boundary Disputes

Guyana was in border disputes with both Suriname, which claimed the land east of the Corentyne River in southeastern Guyana, and Venezuela which claims the land west of the Essequibo River as part of Guyana Essequibo. The maritime component of the territorial dispute with Suriname was arbitrated by the United Nations Convention on Law of the Sea, and a ruling was announced in September 21, 2007. The ruling concerning the Caribbean Sea north of both nations found both parties violated treaty obligations and declined to order any compensation to either party.

When the British surveyed British Guiana in 1840, they included the entire Cuyuni River basin within the colony. Venezuela did not agree with this as it claimed all lands west of the Essequibo River. In 1898, at Venezuela's request, an international arbitration tribunal was convened, and in 1899 they issued an award giving about 94% of the disputed territory to British Guiana.

Venezuela and Great Britain accepted the award by treaty in 1905, but Venezuela raised the issue again at the time of Guyana's independence and continues to claim Guayana Esequiba. Venezuela calls this region "Zona en Reclamation" (Reclamation Zone), and Venezuelan maps of the national territory routinely include it, drawing it in with dashed lines.

Specific small disputed areas involving Guyana are Ankoko Island with Venezuela; Corentyne River with Suriname; and New River Triangle with Suriname.

Environment and Biodiversity

The following habitats have been categorised for Guyana: coastal, marine, littoral, estuarine palustrine, mangrove, riverine, lacustrine, swamp, savanna, white sand forest, brown sand forest, montane, cloud forest, moist lowland and dry evergreen scrub forests (NBAP, 1999). About 14 areas of biological interest have been identified as possible hotspots for a National Protected Area System.

More than 80% of Guyana is still covered by forests, ranging from dry evergreen and seasonal forests to montane and lowland evergreen rain forests. These forests are home to more than a thousand species of trees. Guyana's tropical climate, unique geology, and relatively pristine ecosystems support extensive areas of species-rich rain forests and natural habitats with high levels of endemism. Approximately eight thousand species of plants occur in Guyana, half of which are found nowhere else.

Guyana is one of the countries with the highest biodiversity in the world. Guyana, with 1,168 vertebrate species, 1,600 bird species, boasts one of the richest mammalian fauna assemblages of any comparably sized area in the world. The Guiana Shield region is little known and extremely rich biologically. Unlike other areas of South America, over 70% of the natural habitat remains pristine.

The rich natural history of British Guiana was described by early explorers Sir Walter Raleigh and Charles Waterton and later by naturalists Sir David Attenborough and Gerald Durrell.

In February 2004, the Government of Guyana issued a title to more than 1 million acres of land in the Konashen Indigenous District declaring this land as the Konashen Community-Owned Conservation Area (COCA), to be managed by the Wai Wai. In doing so Guyana created the world's largest Community-Owned Conservation Area.

This important event followed a request made by the Wai Wai community to the government of Guyana and Conservation International Guyana (CIG) for assistance in developing a sustainable plan for their lands in Konashen. The three parties signed a Memorandum of Cooperation which outlines a plan for sustainable use of the Konashen COCA's biological resources, identifies threats to the area's biodiversity, and helps develop projects to increase awareness of the COCA as well as generate the income necessary to maintain its protected status.

The Konashen Indigenous District of Southern Guyana houses the headwaters of the Essequibo River, Guyana's principal water source, and drains the Kassikaityu, Kamoa, Sipu and Chodikar rivers. Southern Guyana is host to some of the most pristine expanses of evergreen forests in the northern part of South America.

Most of the forests found here are tall, evergreen hill-land and lower montane forests, with large expanses of flooded forest along major rivers. Thanks to the very low human population density of the area, most of these forests are still intact. The Smithsonian Institution has identified nearly 2,700 species of plants from this region, representing 239 distinct families, and there are certainly additional species still to be recorded.

Such incredible diversity of plants supports even more impressive diversity of animal life, recently documented by a biological survey organised by Conservation International. The clean, unpolluted waters of the Essequibo watershed support a remarkable diversity of fish and aquatic invertebrates, and are home to giant river otters, capybaras, and several species of caimans.

On land, large mammals, such as jaguars, tapirs, bush dogs, giant anteaters, and saki monkeys are still common. Over 400 species of birds have been reported from the region, and the reptile and amphibian faunas are similarly rich. The Konashen COCA forests are also home to countless species of insects, arachnids, and other invertebrates, many of which are still undiscovered and unnamed.

The Konashen COCA is relatively unique in that it contains a high level of biological diversity and richness that remains in nearly pristine condition; such places have become rare on earth. This fact has given rise to various non-exploitative, environmentally sustainable industries such as ecotourism, successfully capitalizing on the biological wealth of the Konashen COCA with comparatively little enduring impact.

World Heritage Site Status

Many countries interested in the conservation and protection of natural and cultural heritage sites of the world accede to the Convention Concerning the Protection of the World Cultural and Natural Heritage that was adopted by UNESCO in 1972. Guyana signed the treaty in 1977, the first Caribbean State Party to do so. In the mid-1990s, Guyana seriously began the process of selecting sites for World Heritage nomination, and three sites were considered: Kaieteur National Park, Shell Beach and Historic Georgetown. By 1997, work on Kaieteur National Park was started, and in 1998 work on Historic Georgetown was begun. To date, however, Guyana has not made a successful nomination.

Guyana submitted the Kaieteur National Park, including the Kaieteur Falls, to UNESCO as its first World Heritage Site nomination. The proposed area and surrounds have some of Guyana's most diversified life zones with one of the highest levels of endemic species found anywhere in South America. The Kaieteur Falls is the most spectacular feature of the park, falling a distance of 226 metres. The nomination of Kaieteur Park as a World Heritage Site was not successful, primarily because the area was seen by the evaluators as being too small, especially when compared with the Central Suriname Nature Reserve that had just been nominated as a World Heritage Site (2000). The dossier was thus returned to Guyana for revision.

Guyana continues in its bid for a World Heritage Site. Work continues, after a period of hiatus, on the nomination dossier for Historic Georgetown. A Tentative List indicating an intention to nominate Historic Georgetown was submitted to UNESCO in December 2004. There is now a small committee put together by the Guyana National Commission for UNESCO to complete the nomination dossier and the management plan for the site. In April 2005, two Dutch experts in conservation spent two weeks in Georgetown supervising architecture staff and students of the University of Guyana in a historic building survey of the selected area. This is part of the data collection for the nomination dossier.

Meanwhile, as a result of the Kaieteur National Park being considered too small, there is a proposal to prepare a nomination for a Cluster Site that will include the Kaieteur National Park, the Iwokrama Forest and the Kanuku Mountains. The Iwokrama Rain Forest, an area rich in biological diversity, has been described by Major General (Retired) Joseph Singh as "a flagship project for conservation." The Kanuku Mountains area is in a pristine state and is home to more than four hundred species of birds and other animals.

There is much work to be done for the successful nomination of these sites to the World Heritage List. The state, the private sector and the ordinary Guyanese citizens each have a role to play in this process and in the later protection of the sites. Inscription on the UNESCO World Heritage will open Guyana to more serious tourists thereby assisting in its economic development.

Guyana exhibits two of the World Wildlife Fund's Global 200 eco-regions most crucial to the conservation of global biodiversity, Guianan moist forests and Guiana Highlands moist forests and is home to several endemic species including the tropical hardwood Greenheart.

Landmarks

St. George's Anglican Cathedral: One of the tallest wooden church structures in the world and the second tallest wooden house of worship after the Todaiji Temple in Japan.

Demerara Harbour Bridge : The world's fourth-longest floating bridge.

Kaieteur Falls.

Caribbean Community (CARICOM) Building: Houses the largest and most powerful economic union in the Caribbean.

Providence Stadium : Situated in Providence on the east bank of the Demerara River and built in time for the ICC World Cup 2007, it is the largest sports stadium in the country. It is also near the Providence Mall, forming a major spot for leisure in Guyana.

Guyana International Conference Centre: Presented as a gift from the People's Republic of China to the Government of Guyana. It is the only one of its kind in the country.

Stabroek Market: A large cast-iron colonial structure that looked like a statue was located next to the Demerara River.

The City Hall: A beautiful wooden structure also from the colonial era.

Queen's College: Top educational institution in Guyana

Economy

The main economic activities in Guyana are agriculture (production of rice and Demerara sugar), bauxite mining, gold mining, timber, shrimp fishing and minerals. Chronic problems include a shortage of skilled labour and a deficient infrastructure. In 2008, the economy witnessed a 3% increase in growth amid the global economic crisis and is expected to grow further in 2009.

Until recently, the government was juggling a sizable external debt against the urgent need for expanded public investment. Low prices for key mining and agricultural commodities combined with troubles in the bauxite and sugar industries had threatened the government's tenuous fiscal position and dimmed prospects for the future. However, the Guyanese economy has rebounded slightly and exhibited moderate economic growth since 1999, thanks to an expansion in the agricultural and mining sectors, a more favourable atmosphere for business initiatives, a more realistic exchange rate, fairly low inflation, and the continued support of international organizations.

The sugar industry, which accounts for 28% of all export earnings, is largely run by the company Guysuco, which employs more people than any other industry. Many industries have a large foreign investment. For example, the mineral industry is heavily invested in by the American company Reynolds Metals and the Canadian Rio Tinto Alcan; the Korean/Malaysian Barama Company has a large stake in the logging industry.

The production of balata (natural latex) was once big business in Guyana. Most of the balata bleeding in Guyana took place in the foothills of the Kanuku Mountains in the Rupununi. Early exploitation also took place in the North West District, but most of the trees in the area were destroyed by illicit bleeding methods that involved cutting down the trees rather than making incisions in them. Uses of balata included the making of cricket balls, the temporary filling of troublesome tooth cavities, and the crafting of figurines and other decorative items (particularly by the Macushi people of the Kanuku mountains).

Major private sector organizations include the Private Sector Commission (PSC) and the Georgetown Chamber of Commerce & Industry (GCCI);

The government initiated a major overhaul of the tax code in early 2007. The Value Added Tax (VAT) was brought into effect, replacing six different taxes. Prior to the implementation of the VAT, it had been relatively easy to evade sales tax, and many businesses were in violation of tax code. Many businesses were very opposed to VAT introduction because of the extra paperwork required; however, the Government has remained firm on the VAT. By replacing several taxes with one flat tax rate, it will also be easier for government auditors to spot embezzlement. While the adjustment to VAT has been difficult, it may improve day-to-day life because of the significant additional funds the government will have available for public spending.

President Bharrat Jagdeo has made debt relief a foremost priority of his administration. He has been quite successful, getting US$800 million of debt written off by the International Monetary Fund (IMF), the World Bank and the Inter-American Development Bank (IDB), in addition to millions more from other industrial nations. Mr. Jagdeo was lauded by IDB President Moreno for his strong leadership and negotiating skills in pursuing debt relief for Guyana and several other regional countries.

Cost of Living

The cost of living in Guyana is high. This is because most of the items used in daily life are imported with high transportation costs involved. Monopoly in some business sectors also causes higher profit booking and further raising of prices. For example, approximate prices (as of January, 2010) of Gasoline (Petrol) is 5 US$ per gallon, and electricity prices are close to 0.33 US$ per unit. A domestic gas bottle (or gas cylinder) is slightly over 20 US$. Rent for average family accommodation may exceed 500 US$ per month in safe urban locations, and personal income tax, which is 33.33% (one third) of total taxable income makes the cost of living higher. An employee's salary is normally paid in Guyanese dollars (1 US Dollar = 205 Guyanese Dollars approx.) and income tax is deducted by the employer.

Demographics

The population of Guyana is approximately 770,000, of which 90% reside on the narrow coastal strip (approximately 10% of the total land area of Guyana). Guyana's coastal strip ranges from between 10 to 40 miles (16 to 64 km) in width.

The present population of Guyana is racially and ethnically heterogeneous, composed chiefly of the descendants of immigrants who came to the country either enslaved, mostly from Africa, or indentured labourers from India, Ireland, the Netherlands, Germany and Portugal.

The population therefore is made up of groups with ethnic backgrounds from India, Africa, Europe, China, with Aboriginal. These groups of diverse nationality backgrounds have been fused together by a common language, i.e., English and Creole. There has been racial tension between the Indo-Caribbean people and the African-Caribbeans.

The largest ethnic sub-group is that of the descendants of immigrants from India also known as East Indians (Indo-Guyanese), comprising 43.5% of the population in 2002. They are followed by people of African heritage (Afro-Guyanese) (30.2%). The third in number are those of mixed heritage (16.7%), while Aboriginals (Arawak, Wai Wai, Carib, Akawaio, Arecuna, Patamona, Wapixana, Macushi and Warao) are fourth making up close to 10% of the population. The smallest groups are the Europeans (including Portuguese), who number at 1,600 individuals, and the Chinese, who number at 1,400 persons. A small group (fewer than 1%) were unable to be classified.

The population distribution in 2002 was determined by ethnic background. The distribution pattern has been similar to those of the 1980 and 1991 censuses, but the share of the two main groups has declined. The East Indians made up 51.9% of the total population in 1980, but by 1991 this had fallen to 48.6%, and then to 43.5% in the 2002 census. Those of African descent increased slightly from 30.8% to 32.3% during the first period (1980 and 1991) before falling to 30.2% in the 2002 census. With small growth in the population, the decline in the shares of the two larger groups has resulted in the relative shares of the multiracial and Amerindian groups.

The Amerindian population rose by 22,097 people between 1991 and 2002. This represents an increase of 47.3% or annual growth of 3.5%. Similarly, the multiracial population increased by 37,788 persons, representing a 43.0% increase or annual growth rate of 3.2% from the base period of 1991 census.

The European and Chinese populations which declined between 1980 and 1991 regained in numbers by the 2002 census by 54.4% (168 persons) and 8.1% (105 persons) respectively. However, because of their relatively small sizes, the increase has little effect on the overall change. The number of Portuguese (4.3% of the population in 1891) has been declining constantly over the decades.

Most Indo-Guyanese are descended from Bhojpuri-speaking Bihari and Uttar Pradesh migrants.

Language

English is the official language of Guyana and used, for example, in its schools. In addition, Cariban languages (Akawaio, Wai-Wai, Arawak and Macushi) are spoken by a small minority, while Guyanese Creole (an English-based creole with African syntax whose grammar is not standardised.) is widely spoken.

Religion

Data from a 2002 census on religious affiliation indicates that approximately 57% of the population are Christian (of those, 17% are Pentecostal, 8% are Roman Catholic, 7% are Anglican, 5% are Seventh-day Adventist, and 20% belong to other Christian denominations). Approximately 28% are Hindu, 7% are Muslim (mostly Sunni), and 2% practice other beliefs (such as the Rastafarian and Bahai faiths). An estimated 4% of the population do not profess any religion.

Most Guyanese Christians are either Protestants or Roman Catholics and include a mix of all races. Hinduism is dominated by the Indians who came to the country in the early 19th century, while Islam varies between the Afro-Guyanese, and Indian-Guyanese.

Government and Politics

Politics of Guyana takes place in a framework of a semi-presidential representative democratic republic, whereby the President of Guyana is the head of government, and of a multi-party system. Executive power is exercised by the government. Legislative power is vested in both the government and the National Assembly of Guyana. Historically, politics are a source of tension in the country, and violent riots have often broken out during elections. During the 1970s and 1980s, the political landscape was dominated by the People's National Congress.

In 1992, the first "free and fair" elections were overseen by former United States President Jimmy Carter, and the People's Progressive Party has led the country since. The two parties are principally organised along ethnic lines and as a result often clash on issues related to the allocation of resources.

Military

The military of Guyana consists of the Guyana Defence Force (GDF), which includes Ground Forces, Coast Guard, and Air Corps. 155,058 males are fit for service (2002 estimates) The Guyana People's Militia and the Guyana National Service are defunct.

Infrastructure, Communications and Health

Transport

There are a total of 116 miles (187 km) of railway, all dedicated to ore transport. There are 4,952 miles (7,970 km) of highway, of which 367 miles (590 km) are paved. Navigable waterways extend to 669 miles (1,077 km), including the Berbice, Demerara, and Essequibo rivers. There are ports at Georgetown, Port Kaituma, and New Amsterdam. There is 1 international airport (Cheddi Jagan International Airport, Timehri); 1 regional airport (Ogle Airport); and about 90 airstrips, 9 of which have paved runways. Guyana and Suriname are the only two countries on the American mainland which drive on the left.

Electricity

The electricity sector in Guyana is dominated by gbvfd, the state-owned vertically integrated utility. Although the country has a large potential for hydroelectric and bagasse-fueled power generation, most of its 226 MW of installed capacity correspond to inefficient thermoelectric diesel-engine driven generators.

Water Supply and Sanitation

Key issues in the water and sanitation sector in Guyana are poor service quality, a low level of cost recovery and low levels of access. A high-profile management contract with the British company Severn Trent was cancelled by the government in February 2007. In 2008 the public utility Guyana Water Inc implemented a Turnaround Plan (TAP) to reduce non-revenue water and to financially consolidate the utility. NRW reduction is expected to be 5% per annum for the three-year period of the plan, A mid term review is now due to examine the success of the TAP.

Communications

Telephone System

Telephones : 110,120 main telephone lines (2005)

Telephones-mobile cellular: 281,400 (2005)

General Assessment: fair system for long-distance service

Domestic: microwave radio relay network for trunk lines; fixed-line teledensity is about 15 per 100 persons; many areas still lack fixed-line telephone services; mobile-cellular teledensity reached 37 per 100 persons in 2005 international: country code-592; tropospheric scatter to Trinidad; satellite earth station-1 Intelsat (Atlantic Ocean)

Radio Broadcast Stations

AM 3, FM 3, shortwave 1 (1998).

Television Broadcast Stations

Television broadcast was officially introduced to Guyana in 1991.

15 (1 public station (channel 11); 14 private stations which relay on US satellite services) (1997)

Internet System

Internet Country Code: .gy

Internet Hosts: 6,218 (2008)

Internet users: 190,000 (2007).

Public Health

Service Delivery

The delivery of health services is provided at five different levels in the public sector:

Level I: Local Health Posts (166 in total) that provide preventive and simple curative care for common diseases and attempt to promote proper health practices. Community health workers staff them.

Level II: Health Centres (109 in total) that provide preventive and rehabilitative care and promotion activities. These are ideally staffed with a medical extension worker or public health nurse, along with a nursing assistant, a dental nurse and a midwife.

Level III: Nineteen District Hospitals (with 473 beds) that provide basic in-patient and outpatient care (although more the latter than the former) and selected diagnostic services. They are also meant to be equipped to provide simple radiological and laboratory services, and to be capable of gynecology, providing preventive and curative dental care. They are designed to serve geographical areas with populations of 10,000 or more.

Level IV: Four Regional Hospitals (with 620 beds) that provide emergency services, routine surgery and obstetrical and gynecological care, dental services, diagnostic services and specialist services in general medicine and pediatrics. They are designed to include the necessary support for this level of medical service in terms of laboratory and X-ray facilities, pharmacies and dietetic expertise. These hospitals are located in Regions 2, 3, 6 and 10.

Level V: The National Referral Hospital (937 beds) in Georgetown that provides a wider range of diagnostic and specialist services, on both an in-patient and out-patient basis; the Psychiatric Hospital in Canje; and the Geriatric Hospital in Georgetown. There is also one children's rehabilitation centre.

This system is structured so that its proper functioning depends intimately on a process of referrals. Except for serious emergencies, patients are to be seen first at the lower levels, and those with problems that cannot be treated at those levels are referred to higher levels in the system. However, in practice, many patients by-pass the lower levels. The health sector is currently unable to offer certain sophisticated tertiary services and specialised medical services, the technology for which is unaffordable in Guyana, or for which the required medical specialists are not available. Even with substantial improvements in the health sector, the need for overseas treatment for some services might remain. The Ministry of Health provides financial assistance to patients requiring such treatment, priority being given to children whose condition can be rehabilitated with significant improvements to their quality of life.

There are 10 hospitals belonging to the private sector and to public corporations, plus diagnostic facilities, clinics and dispensaries in those sectors. These ten hospitals provide for 548 beds. Eighteen clinics and dispensaries are owned by GUYSUCO.

The Ministry of Health and Labour is responsible for the funding of the National Referral Hospital in Georgetown, which has recently been made a public corporation managed by an independent Board. Region 6 is responsible for the management of the National Psychiatric Hospital. The Geriatric Hospital, previously administered by the Ministry of Labour, became the responsibility of the Ministry of Human Resources and Social Security in December 1997.

Health Conditions

One of the most unfortunate consequences of Guyana's economic decline in the 1970s and 1980s was that it led to very poor health conditions for a large part of the population. Basic health services in the interior are primitive to non-existent, and some procedures are not available at all. The US State Department Consular Information Sheet warns "Medical care is available for minor medical conditions.

Emergency care and hospitalization for major medical illnesses or surgery is limited, because of a lack of appropriately trained specialists, below standard in-hospital care, and poor sanitation. Ambulance service is substandard and may not routinely be available for emergencies." Many Guyanese seek medical care in the United States, Trinidad and Tobago or Cuba.

Compared with other neighbouring countries, Guyana ranks poorly in regard to basic health indicators. In 1998, life expectancy at birth was estimated at 66.0 years for Guyana, which is much less than surrounding countries. The infant mortality rate in 1998 was 24.2. Maternal mortality rates in Guyana are also relatively high, being estimated at 124.6/1000 for 1998. Although Guyana's health profile falls short in comparison with many of its Caribbean neighbours, there has been remarkable progress since 1988, and the Ministry of Health is working to upgrade conditions, procedures, and facilities.

The leading causes of mortality for all age groups are cerebrovascular diseases (11.6%); ischemic heart disease (9.9%); immunity disorders (7.1%); diseases of the respiratory system (6.8%); diseases of pulmonary circulation and other forms of heart disease (6.6%); endocrine and metabolic diseases (5.5%); diseases of other parts of the Digestive System (5.2%); violence (5.1%); certain condition originating in the prenatal period (4.3%); and hypertensive diseases (3.9%). The ten leading causes of morbidity for all age groups are, in decreasing order: malaria; acute respiratory infections; symptoms, signs and ill defined or unknown conditions; hypertension; accident and injuries; acute diarrhoeal disease; diabetes mellitus; worm infestation; rheumatic arthritis; and mental and nervous disorders.

This morbidity profile indicates that it can be improved substantially through enhanced preventive health care, better education on health issues, more widespread access to potable water and sanitation services, and increased access to basic health care of good quality. A number of non-governmental organisations, including Health and Educational Relief for Guyana (HERG, INC) and Guyana Medical Relief (GMR, INC) are currently working to address these issues by improving healthcare access and educational infrastructure.

Guyana has experienced an upswing in violent crime and homicide in 2008 while the numbers of murders reported actually dropped in 2007 over the previous few years, with a murder rate of 15.1 people for each 100,000, in contrast to 2008 (up to the end of July) that number has risen to 26 per 100,000 similar to the rate experienced in 2003. Guyana suffers from the highest suicide rate of any South American country. Guyana Health

Minister Leslie Ramsammy estimates that at least 200 people commit suicide each year in Guyana, or 27.2 people for each 100,000 people each year.

Education

Guyana's educational system was at one time considered to be among the best in the Caribbean, but it significantly deteriorated in the 1980s because of the emigration of highly educated citizens and the lack of appropriate funding. Although the education system has recovered somewhat in the 1990s, it still does not produce the quality of educated students necessary for Guyana to modernise its workforce. The country lacks a critical mass of expertise in many of the disciplines and activities on which it depends.

The educational system does not sufficiently focus on the training of Guyanese in science and technology, technical and vocational subjects, business management, nor computer sciences. The Guyanese education system is modelled after the former British education system. Students are expected to write SSEE (secondary school entrance exam) by grade 6 for entrance into High School in grade 7. They write CXC at the end of high school. Recently they have introduced the CAPE exams which all other Caribbean countries have introduced. The A-level system left over from the British era has all but disappeared and is offered only in a few schools.

Further adding to the problems of the educational system, many of the better-educated professional teachers have emigrated to other countries over the past two decades, mainly because of low pay, lack of opportunities and crime. As a result, there is a lack of trained teachers at every level of Guyana's educational system. There are however several very good Private schools that have sprung up over the last fifteen years. Those schools offer a varied and balanced curriculum. However, the top government schools have nonetheless continued their dominance in academic performance outshining these private schools over the years.

Culture

Guyana, along with Suriname, French Guiana, and Brazil, is one of the four non-Hispanic nations in South America.

Guyana's culture is very similar to that of the English-speaking Caribbean, and has historically been tied to the English speaking Caribbean as part of the British Empire when it became a possession in the nineteenth century. Guyana is a founding member of the Caricom (Caribbean Community) economic bloc and also the home of the Bloc's Headquarters, the CARICOM Secretariat.

Guyana's geographical location, its sparsely populated rain forest regions, and its substantial Amerindian population differentiate it from English-speaking Caribbean countries. Its blend of Indo-Guyanese (East Indian) and Afro-Guyanese (African) cultures gives it similarities to Trinidad and distinguishes it from other parts of the Americas. Guyana shares similar interests with the islands in the West Indies, such as food, festive events, music, sports, etc.

Guyana plays international cricket as a part of the West Indies cricket team, and the Guyana team plays first class cricket against other nations of the Caribbean. In March and April 2007 Guyana co-hosted the Cricket World Cup 2007. In addition to its CARICOM membership, Guyana is a member of CONCACAF, the international football federation for North and Central America and the Caribbean.

Events include Mashramani (Mash), Phagwah (Holi), and Deepavali (Diwali).

Sports

The major sports in Guyana are cricket (Guyana is part of the West Indies as defined for international cricket purposes), softball cricket (beach cricket) and football (soccer). Minor sports include netball, rounders, lawn tennis, basketball, table tennis, boxing, squash and a few others.

Guyana played host to international cricket matches as part of the 2007 Cricket World Cup. The new 15,000-seat Providence Stadium, also referred to as Guyana National Stadium, was built in time for the World Cup and was ready for the beginning of play on March 28. At the first international game of CWC 2007 at the stadium, Lasith Malinga of the Sri Lankan team took four wickets in four consecutive deliveries.

Bibliography

Belden C.: *Landscapes of the Sacred: Geography and Narrative in American Spirituality*, New York: Paulist Press, 1988.

Berry, J.K. : *Beyond Mapping, Concepts, Algorithms and Issues in GIS*, Fort Collins, CO: GIS World Books, 1993.

Bolstad, P. : *GIS Fundamentals, A first text on Geographic Information Systems, Second Edition*, White Bear Lake, MN, Eider Press, 2005.

Cronon, William: *Changes in the Land: Indians, Colonists, and the Ecology of New England.* New York: Hill and Wang, 1983.

Davis, Mike: *Ecology of Fear: Los Angeles and the Imagination of Disaster*. New York: Vintage, 1999.

Elangovan K. : *GIS, Fundamentals, Applications and Population Geography*, New India Publishing Agency, New Delhi 2006.

Harvey, David: *Explanation in Geography*. London: Edward Arnold, 1969.

Harvey, Francis : *A Primer of GIS, Fundamental Geographic and Cartographic Concepts,* The Guilford Press, 2008.

Jackson, Peter: *Maps of Meaning: An Introduction to Cultural Geography*. London and Boston: Unwin Hyman, 1989.

James S. Duncan: *Writing Worlds: Discourse, Text, and Metaphor in the Representation of Landscape*. London and New York: Routledge, 1992.

James, T.: *The Best Poor Man's Country: A Geographical Study of Early Southeastern Pennsylvania*. Baltimore: Johns Hopkins Press, 1972.

King, Anthony D.: *Buildings and Society: Essays on the Social Development of the Built Enviroment*. London: Routledge & Kegan Paul, 1980.

Lewis, Peirce. *New Orleans: The Making of an Urban Landscape*. Cambridge, MA: Ballinger Publishing Co., 1976. Annotation

Liam Kennedy: *Urban Space and Representation*. Sterling, VA: Pluto Press, 2000.

Lung, Y., and van Tulder, R. : *Cars, Carriers of Regionalism*, Palgrave Macmillan, New York, 2004.

McDonnell, R.A. : *Principles of Geographical Information Systems*, Oxford University Press, Oxford, 1998.

Michael P.: *The Making of the American Landscape*. Boston: Unwin Hyman, 1990.

Michael, H.: *City Form and Natural Process: Towards a New Urban Vernacular*. London: Croom Helm, 1984.

Peter, G.: *The Slow Plague: A Geography of the AIDS Pandemic*. Oxford and Cambridge, MA: Blackwell Publishers, 1993.

Ray Allen: *The Far Western Frontier, 1830-1860*. The New American Nation Series. New York: Harper, 1956.

Richard SCASE : *Managing Creativity,* Open University Press, 2000.

Stephen, D.: *Fields of Vision: Landscape Imagery and National Identity in England and the United States*. Cambridge: Polity, 1993.

Tulder, R. : *Cars, Carriers of Regionalism*, Palgrave Macmillan, New York, 2004.

Victor, A. : *Geography of Art and Culture*, Contributions to Economic Analysis, Elsevier, 2004.

Watkins, Eric: *The Middle Eastern Environment*, London, The British Society for Middle Eastern Studie, 1995.

William G. BOWEN : *Performing Arts, The Economic Dilemma,* Thousand Oaks, SAGE, London, 2002.

Worboys, Michael, and Matt Duckham. : *GIS, a Computing Perspective*, Boca Raton, CRC Press, 2004.

Index

□□□